Quantum Narratives

ALSO BY STEPHANIE RIGGS

The End of Storytelling

Quantum Narratives

STEPHANIE RIGGS

BEAT MEDIA GROUP

BEAT MEDIA GROUP

Copyright © 2025 by Stephanie Riggs

All rights reserved. No part of this publication may be reproduced, distributed, or transmitted in any form or by any means, including photocopying, recording, or other electronic or mechanical methods, without the prior written permission of the publisher, except in the case of brief quotations embodied in critical reviews and certain other noncommercial uses permitted by copyright law.

Book design and illustrations by Maya P. Lim, mayaplim.com

First print edition March 2025

Hardcover ISBN 978-1-7329559-4-3

Published by Beat Media Group

quantumnarratives.com

For Charlie & Max
the paradox of duality at the center of my universe

ACKNOWLEDGMENTS

To my family—thank you for your unwavering support and patient love as these ideas took shape, unraveled, and reassembled over the years. If time is truly relative, then the late nights and moments of doubt must have stretched far longer for you than they did for me.

To all my colleagues—thank you for the many invigorating conversations, long debates, and sparks of inspiration along the way. Blair Erickson, Jackson Gallagher, Tommy Honton and James Wallman, you especially listened when the ideas were most raw, never questioning whether they would one day take shape. My early draft readers—Rob Auten, Justin Bolognino, David Shiyang Liu, Christopher Morrison, Kathryn McElroy, and Evan True—bravely waded through tangled thoughts and half-formed arguments, offering sharp, candid insights and encouragement at critical moments. Dana Buning, you have been my guardian spirit on this journey, grounding me and guiding me. I am forever grateful. The steady hand of my editor, Sarah Haskins, and the remarkable intuition and talent of Maya P. Lim—who not only designs lines beautifully but seems to read between them with uncanny precision—ultimately helped this book find its final form.

And to the pioneers in physics, computer science, psychology, gaming, storytelling, and beyond—you turned thought experiments into theories, equations into inventions, and ideas into worlds. This book is, in many ways, a continuation of the paths you forged, though, as Einstein reminded us, time is an illusion, and I like to think we are all walking this path together.

CONTENTS

Preface	i
Setting the Stage	1
A Narrative Paradox Emerges	15
Classical Catastrophes	33
Quantum Narratives	53
The New Domain	85
Epilogue: A Play on Postulates	97
Notes	110
Index	116

PREFACE

For decades it has seemed like an exercise in futility to balance interactivity, which I first deeply explored in the medium of virtual reality, with the constraints of linear storytelling using the coding capabilities available. Even with today's newest technologies, the answer has always felt like it needed to be holistic. It wasn't just computer systems that needed to be investigated; we needed to critically evaluate the foundations and experience of story as well.

In my previous book, *The End of Storytelling*, I challenged readers to rethink assumptions entrenched in developing content for framed, linear mediums like film, television, and theater and explore new techniques for narratives in immersive and interactive mediums. Central to this exploration is the concept of the Storyplex, a paradigm for conceptualizing new kinds of narratives in emerging technologies. I defined the Storyplex as a dynamic network that interweaves story, technology, and humanity. The original draft of that book included a chapter on Quantum Narratives. I knew where I was headed with my research at the time and decided to omit the chapter to keep the content more accessible. In retrospect, this decision may have left certain aspects of the Storyplex concept underexplored, based on the questions I received in lectures and talks on *The End of Storytelling* afterwards. I hope this book will bring clarity to those readers.

On the positive side, extracting the Quantum Narratives section gave me more time to research, explore, test, and evolve my early framework. It also allowed time for interactive technology, game engines, and artificial intelligence to become more powerful and ubiquitous. Conversations on generative content and spatial stories that interested only a handful of people seven years ago now have much greater visibility. The deeper I went into the parallels between scientific thought and what I view as a necessary shift in narrative paradigm, the more they reinforced each other.

To immerse myself in the world of the natural sciences, I reached out to colleagues with PhDs in physics, molecular biology, and mathematics. They kindly directed me to materials where I could dive into the world of physics, which led me to quantum mechanics. I am indebted to the many resources now available that explain the complexities of the mathematics in understandable terms. My intention was never to become a physicist, but to glide along the surface of comprehension far enough that I could determine whether or not there were logical commonalities between physics and the kind of story I have dreamed about since I first put on a VR headset. The rabbit hole went deep and, as the parallels and overlaps made themselves apparent, I went happily with it.

PREFACE

Applying a term like "quantum" to the field of narrative is loaded. Physicists, chemists, engineers, computer scientists, mathematicians, philosophers, astronomers, and contributors from many other fields have dedicated decades and even lifetimes contributing to a science that is abstract, counterintuitive, and mathematically complex. What follows is not intended to be a one-to-one correlation between the mathematically provable universe and the mysteries of humanity.

On the topic of terms, the words "story" and "narrative" are often used interchangeably, but I contend that there is a distinction. Story is the content being conveyed. Narrative is how the content is expressed and interpreted. We apply different narratives to the same story all the time. The same story can be told from various perspectives such as first-person, third-person, or nonlinear with flashbacks or flashforwards. In essence, the story is the sequence of events and the characters involved, while the narrative is the manner in which these events and characters are presented to the observer and their experience of the story. In this way, narrative is an evolution of story.

In this book I condense my learnings and observations. Once again, there are things I have chosen to omit for the sake of brevity. The importance of one thing became clear to me in the course of this exploration that I can't leave out: as new technologies come along, what is more important than the tools is *us*—how we think about them and how we use them; the technologies we build and the stories we share are as much a reflection of our humanity as our ingenuity.

Stephanie Riggs
Brooklyn, NY
October 2024

Nature isn't classical, dammit, and if you want to make a simulation of nature, you'd better make it quantum mechanical.
—RICHARD FEYNMAN

Every act of perception is to some degree an act of creation, and every act of memory is to some degree an act of imagination.
—OLIVER SACKS

In the end, we all become stories.
—MARGARET ATWOOD

SETTING THE STAGE

In a dimly lit movie theater, the dramatic climax of a slowly building adventure film explodes onto the screen. A flash of lightning illuminates the audience's emotionally wrought faces for a fraction of a second. The heroine stands at a crossroads, eyes darting across the barren landscape, wracked with inner turmoil. Every muscle is taut, every vein pulsing with adrenaline. The music violently crescendos, building to a fever pitch. Just as she opens her mouth to announce the decision that will determine her fate…the screen abruptly goes black. No resolution. No closure. The audience is left hanging, on the edge of their seats with clenched fists and bated breath. It's maddeningly unsatisfying.

Across Western cultures and throughout history, we've been told that stories are supposed to have a protagonist undergoing transformation in a well-designed plot written by an author. That's the unspoken contract between storyteller and audience.

But what if that contract is being rewritten? What if the stories that stay with us, the ones we think about long after the credits roll or the book closes, are the ones that we are invited to not only see ourselves in, but be a part of?

As technologies expand the possibilities of our reality and our stories, as our content becomes increasingly interactive, as narratives resist the old rules of linearity, must we still insist on using ancient storytelling frameworks to engage audiences? Or, are we ready to embrace narratives that dynamically live, respond, and evolve—as unpredictable as life itself?

POETIC FOUNDATIONS

The Ancient Greek philosopher Aristotle, known for his significant contributions to ethics, logic, physics, and biology, also profoundly influenced Western dramatic theory. His seminal work, *Poetics*, outlined what he believed were crucial elements for creating a compelling drama. Tragedy, he said, is designed to have a serious purpose. The form aims to evoke emotions of pity and fear in the audience, leading to a *catharsis*—the purging of emotion that leads to an uplifting emotional release. To achieve this, he emphasized the necessity of a coherent, well-structured plot.

Effective plots adhere to three main Unities, according to Aristotle. The Unity of Action dictates that a play should have one central storyline, with minimal subplots or distractions. The Unity of Time suggests that the events of the play should take place within a single day. And the Unity of Place means that the setting should remain consistent throughout. By focusing on a singular story in real time, Aristotle believed the emotional impact on the audience is heightened.

These principles mirrored the cultural ideals of his time, where beauty was closely associated with harmony and structured order. Fixed character archetypes—heroes, villains, and tragic figures—ensured predictable roles. Moral absolutes presented clear divisions between good and evil. Strict adherence to genre reinforced this predictability, as each story form (such as tragedy or comedy) came with clear thematic boundaries, tones, and expected outcomes.

In Aristotle's teleological view of the natural world, everything in nature has a purpose and moves toward an end goal. He believed fire rises in order to reach the heavens, while stones fall to seek their "natural place"—the universe's center, which he was certain lay at Earth's core. He saw symmetry and order as fundamental to the cosmos. A central, stationary Earth orbited in perfect circles by the planets aligned with religious and philosophical ideas of a divinely ordered universe, which placed humans in a privileged, central position.

Aristotle's explanation of how and why things exist, known as the Four Causes, once dominated our understanding of the natural world. To grasp this concept, imagine yourself in a gallery in ancient Greece in 335 BCE, during his time. As you stand before a marble statue, the philosopher explains its existence through the Four Causes. First, feel the cool marble under your fingers—the marble is the "material cause," the raw substance of the statue. Then, let your eyes trace the statue's form, its intricate design and shape—this is the "formal cause," the envisioned blueprint that guided its creation. Next, picture the sculptor at work, chisel in hand—the "efficient cause," the force that brought the vision to life. Finally, consider why this statue was made: to honor a god, to mark a victory, or to capture beauty. This is the "final cause," the purpose driving its existence. Imagine for a moment how your worldview might be different if you truly believed, as Aristotle did, that something's existence was defined by its material, formal, efficient, and final cause.

For over 2,000 years the Four Causes were central to Western thought, shaping science, philosophy, and theology. The order and stability of a world seen as fixed and purposeful were reflected in these reasoning models as well as the early rules of storytelling. Yet, as our tools, observations, and beliefs have evolved, so too has our understanding of reality—and our place in it.

THE POWER OF STORY

When Copernicus published his theory of heliocentrism in 1543, he was fully aware it would cause a clash with religious beliefs. And although he supported his claims that the Sun, rather than the Earth, is at the center of the universe with ample observations and precise calculations, he still felt the need to poetically compel the readers of his revolutionary theory through story:

> *"At rest, however, in the middle of everything is the sun. For in this most beautiful temple, who would place this lamp in another or better position than that from which it can light up the whole thing at the same time? For, the sun is not inappropriately called by some people the lantern of the universe, its mind by others, and its ruler by still others. [Hermes] the Thrice Greatest labels it a visible god, and Sophocles' Electra, the all-seeing. Thus indeed, as though seated on a royal throne, the sun governs the family of planets revolving around it."*[1]

Having courted the Catholic Church and dedicated the book of his findings to Pope Paul III with a palatable story, Copernicus escaped condemnation and was allowed to live out his days in peace. But nearly seventy years later, Galileo Galilei, whose magnification improvements on the telescope revealed previously invisible details and ultimately provided even more compelling evidence against the geocentric model, was taken to trial by the Roman Catholic Inquisition. Galileo was found "vehemently suspect of heresy,"[2] forced to recant his views, and put under house arrest for the rest of his life. Galileo's unwavering assertion of the theory also landed Copernicus' writings in the Church's Index of Forbidden Books.

Even when faced with compelling evidence, many people—especially powerful institutions—have found it easier to question the messenger than to embrace a new model that requires profound change. Armed with repeatable, quantifiable, irrefutable evidence, and persecuted for it, the Western thinkers increasingly distanced themselves from divine narratives, both in methodology and language. Sir Isaac Newton's 1687 magnum opus, *Principia Mathematica*, famously formulated the laws of motion and universal gravitation. It was written in Latin, a language in rapid decline outside of academic institutions, and relied heavily on explanations described only through mathematical symbols. The inaccessibility of these languages to the common audience was intentional according to his colleagues.

> "...(Newton) told me, he designedly made his *Principa* abstruse; but yet so as to be understood by able Mathematicians, who he imagined, by comprehending his Demonstrations, would concur with him in his Theory..."[3]

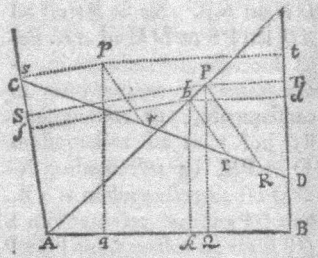

FIGURE 1. Excerpt from *Philosophiae Naturalis Principia Mathematica* (Newton, 1687).[4]

In a calculated maneuver, Newton erected invisible walls of exclusivity around his revolutionary discoveries, effectively barricading them from the public. This further deepened the widening divide between traditional beliefs and scientific thinking.

Albert Einstein's methodical unveiling of General Relativity in 1915 exemplified a more collaborative evolution of revolutionary thought. His groundbreaking theory was not presented in a single submission but rather delivered in four works to the Prussian Academy of Sciences on consecutive Thursdays in November. Initially, he provided a general overview that could be understood by a broader academic audience. The final paper, however, was highly abstract and predominantly written in the language of pure mathematics.

Since the scientific community of the 20th century was far more interconnected and collaborative than during Newton's time, Einstein's peers had varying expertise in mathematics. By

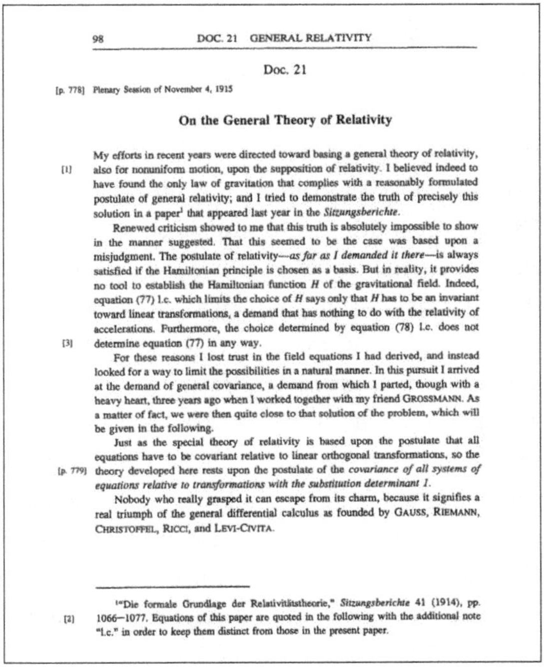

98 DOC. 21 GENERAL RELATIVITY

Doc. 21

[p. 778] Plenary Session of November 4, 1915

On the General Theory of Relativity

My efforts in recent years were directed toward basing a general theory of relativity,
[1] also for nonuniform motion, upon the supposition of relativity. I believed indeed to have found the only law of gravitation that complies with a reasonably formulated postulate of general relativity; and I tried to demonstrate the truth of precisely this solution in a paper[1] that appeared last year in the *Sitzungsberichte*.

Renewed criticism showed to me that this truth is absolutely impossible to show in the manner suggested. That this seemed to be the case was based upon a misjudgment. The postulate of relativity—*as far as I demanded it there*—is always satisfied if the Hamiltonian principle is chosen as a basis. But in reality, it provides no tool to establish the Hamiltonian function H of the gravitational field. Indeed, equation (77) l.c. which limits the choice of H says only that H has to be an invariant toward linear transformations, a demand that has nothing to do with the relativity of accelerations. Furthermore, the choice determined by equation (78) l.c. does not
[3] determine equation (77) in any way.

For these reasons I lost trust in the field equations I had derived, and instead looked for a way to limit the possibilities in a natural manner. In this pursuit I arrived at the demand of general covariance, a demand from which I parted, though with a heavy heart, three years ago when I worked together with my friend GROSSMANN. As a matter of fact, we were then quite close to that solution of the problem, which will be given in the following.

Just as the special theory of relativity is based upon the postulate that all equations have to be covariant relative to linear orthogonal transformations, so the
[p. 779] theory developed here rests upon the postulate of the *covariance of all systems of equations relative to transformations with the substitution determinant 1*.

Nobody who really grasped it can escape from its charm, because it signifies a real triumph of the general differential calculus as founded by GAUSS, RIEMANN, CHRISTOFFEL, RICCI, and LEVI-CIVITA.

[1] "Die formale Grundlage der Relativitätstheorie," *Sitzungsberichte* 41 (1914), pp.
[2] 1066–1077. Equations of this paper are quoted in the following with the additional note "l.c." in order to keep them distinct from those in the present paper.

→
FIGURE 2. Excerpt from *The Collected Papers of Albert Einstein*, English translation (Einstein, 1915).[5]

beginning with conceptual understanding before supporting it with complex equations, he was able to ease the acceptance of his theories. Still, the general public needs a metaphor to grasp concepts like spacetime, which merges space and time into a single continuum. Imagine a trampoline stretched tight; this represents spacetime. When a heavy bowling ball is placed in the center of the trampoline, it creates a dip in the fabric. Now, roll a smaller ball, like a marble, across the trampoline. The marble doesn't move in a straight line—it curves toward the bowling ball because of the dip. This metaphor functions as a micro-story, taking us on a journey of understanding. Using story to communicate complex and abstract ideas has become an important part of scientific communication.

From Aboriginal origin stories explaining the land and our place in it to artificial intelligence's provocation of what it means to be human, we search for understanding and purpose. Stories have long been how we make sense of the unknown. While Aristotle looked for purpose within the framework of the Four Causes, the work of modern theoretical physicist and esteemed author Stephen Hawking centered on understanding the universe through physics and cosmology, particularly in developing a "theory of everything" that might one day illuminate the underlying laws governing existence. Hawking speculated about the origins and nature of the universe, but he expressed the limitations of mathematics in that search: "...the usual approach of science of constructing a mathematical model cannot answer the questions of why there should be a universe for the model to describe."[6] Calculations in science are primed to explain the how and not the why. Stories, on the other hand, tend to explore the why without the how. They have a way of democratizing information. Hawking's *A Brief History of Time*, published in 1988, is one of the most well-known and influential popular science books of all time. In it, his accessible language and use of clear, illustrative explanations empowered readers without a scientific background to engage with profound ideas about our existence. This was as intentional as Newton's use of Latin and complex mathematics, according to Hawking:

> *"...if we do discover a complete unified theory, it should in time be understandable in broad principles by everyone, not just a few scientists. Then we shall all, philosophers, scientists, and just ordinary people, be able to take part in the discussion of the question of why it is that we and the universe exist. If we find the answer to that, it would be the ultimate triumph of human reason—for then we would know the mind of God."* [6]

It seems nothing is more important—or elusive—to us than ourselves and our place in this universe. Our perceived understanding of reality provides us with a sense of control and security. But when confronted with a new paradigm that shatters our sense of certainty, we are faced with the daunting task of abandoning comforting beliefs that once gave us clarity and purpose. As humans, we cling to what we believe to be true, and this preference for the familiar shapes our identities and perception of the world around us. Resisting change is not just limited to individual minds, but also ingrained in social norms, cultural traditions, and institutional structures—all designed to safeguard the entrenched narratives passed down for generations.

PRIMACY OF PLOT AND CHARACTER

If equations are the cornerstone of rigorous science, what are the foundations of a powerful story? According to Aristotle, an effective and engaging story is built upon six core components: plot, character, thought, diction, melody, and spectacle. While originally defined for ancient Greek tragedy, these elements are still structuring stories and genres today.

Aristotle saw the plot, or *mythos*, as the "soul" of the tragedy. As the most crucial element, plot should organize events and actions in a meaningful way that captures the audience's interest and drives the story forward. He noted that plays with a clear beginning, middle, and end were more satisfying to audiences and codified the three Unities of Action, Time, and Place to serve this purpose. His emphasis on plot as the guiding principle reflects his belief that a story's purpose is to evoke emotions through a tightly woven sequence of events.

Many story structures have since expanded on the simplified beginning-middle-end sequence. One of the most widely taught is Freytag's Pyramid. Articulated by Gustav Freytag in 1863, it maps out a five-part structure wherein the action of a story rises and falls in the shape of a pyramid: exposition (introduction), rising action

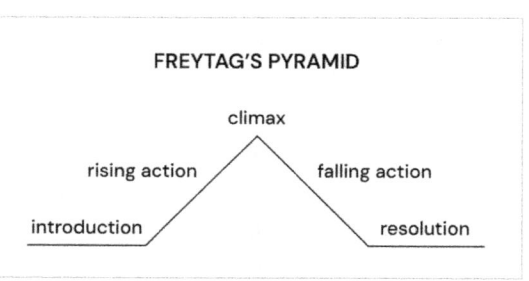

→ FIGURE 3. Diagram of Freytag's Pyramid, illustrating the progression of a story through a five-part dramatic structure.

(building tension), climax (highest point of conflict), falling action (events leading to resolution), and resolution (conclusion). Other structures, like the Hero's Journey popularized by Joseph Campbell and the Three-Act Structure widely used in screenwriting, all provide an organizational method for crafting engaging and cohesive stories by breaking them into specific stages or "beats."

Central to these Western structures is the use of conflict as a driving force in storytelling. While Aristotle focused on plot as a series of causally connected events, later theorists identified conflict as the engine that propels the narrative forward, heightening tension and deepening audience engagement. Freytag placed conflict at the heart of the pyramid's rising action and climax, marking it as essential for driving the stakes and creating catharsis. Similarly, Campbell's Hero's Journey structure revolves around conflicts—both internal and external—that the protagonist must confront and overcome to achieve transformation. By explicitly centering conflict, their structures mirror the psychological dynamics of being the center of our own stories.

We see reflections of our personal struggles, triumphs, and transformations in the characters we encounter. Aristotle noted that audiences connect deeply with characters who exhibit consistent traits and motivations. This, combined with his view of stories as a reflection of universal human experiences, created a focus on a single protagonist undergoing a personal journey that taps into the timeless human desire to find meaning through individual transformation. This power of personal identification was strikingly evident in the public response to Johann Wolfgang Goethe's 1774 novel *The Sorrows of Young Werther*, which sparked "Werther Fever" across Europe as young men began imitating the protagonist's mannerisms, dress style, and, tragically, his suicide.

And while many viewers likely would not condone or relate to Walter White's actions in the series *Breaking Bad*, they can connect with the underlying emotions and motivations driving his choices. His transformation from a mild-mannered chemistry teacher to a ruthless drug lord is driven by his ambition, pride, and desperation, which ultimately leads to his downfall. Our natural connection to Walter's human qualities draws us into his story.

A tragic hero's downfall is not merely a consequence of personal flaws or virtues. From a story standpoint, it is a necessary step toward an ultimate catharsis. Walter's choices and flaws are integral to the narrative, as his eventual demise not only resolves his personal arc but also provides the audience with a sense of catharsis by bringing the story's moral and emotional tensions to a climactic resolution. Ultimately, characters are traditionally vessels through which the plot unfolds, serving as an essential but secondary element in the storytelling process.

Today's linear storytellers often employ multiple intertwined plotlines, span various time periods or even nonlinear timelines, and set their stories across multiple locales or environments—defying Aristotle's prescriptive conventions. Shows like *Westworld* and movies like *Everything, Everywhere, All At Once* weave together disparate stories that span vast timelines and places. Even when narratives defy traditional structure, the characters and their motivations provide a familiar entry point that keeps audiences emotionally engaged.

The purpose of the remaining elements of thought, diction, melody, and spectacle are to support plot and character, according to Aristotle. Thought (or theme) is the underlying message or philosophical idea in the story woven into character dialogue or decision-making. Diction (language and expression) conveys plot and character, while melody (rhythm and musical elements) enhances mood and emotional resonance but doesn't drive the story's core. The visual elements of spectacle are seen as superficial, appealing primarily to the senses rather than to the mind or emotions. Aristotle argued that even without elaborate visuals, a well-crafted story with a strong plot and developed characters could achieve the desired emotional effect of catharsis.

Rooted in observations that resonate deeply with human experience, Aristotle's *Poetics* has survived millennia. His elements of drama reflect the way we, as humans, perceive and interpret stories. Whether in a Greek amphitheater, a Shakespearean playhouse, a Hollywood blockbuster, or a video game, the importance of plot and character in Western drama aligns with our desire to find coherence, grapple with conflict, and connect emotionally. Aristotle's theories reflect us back to ourselves.

But what happens when we want to interact with that reflection?

A NARRATIVE PARADOX EMERGES

As storytellers began to experiment with interactivity, they started with tightly controlled plot points, utilizing branching storylines where readers could make decisions that led to different outcomes. This approach was popularized by the *Choose Your Own Adventure* book series and explored in other forms, including theater. In Ayn Rand's courtroom drama "The Night of January 16th," the jury, made of audience members, determines the show's ending. If the jury finds the suspect guilty, one ending is performed; a different one is performed if they find her innocent.

With the advent of digital technology, interactive possibilities expanded dramatically. The genre of interactive digital storytelling emerged in the late 20th century, evolving from text-based adventure games like *Colossal Cave Adventure* (1976) and *Zork*

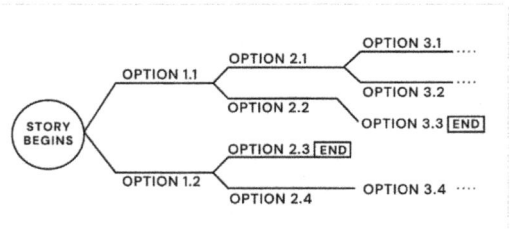

FIGURE 4. Diagram of a branching narrative structure with multiple story paths and potential endings.

(1980) that allowed players to explore storyworlds and solve puzzles in a story through simple commands. Later, games began to incorporate visuals, exemplified by the iconic terrain graphics of *Myst* (1993). For nearly a decade, it was the bestselling computer game in the world—but according to its creator Robyn Miller, "We didn't really tell stories. They were just these worlds that you would explore."[7]

Computer and console games became more ambitious as processing capabilities increased. World-building, character progression, and gameplay mechanics all became points of choice for players. Open-world games such as *The Sims* (2000) offered gameplay where players could create their own stories within a sandbox environment. Game designers started incorporating story to enhance emotional engagement and to make games more than just skill-based challenges. In *Half-Life* (1998), players are guided through a predetermined storyline and have no control over its progression or outcomes. Eventually, games became experiences where players were directly involved in how the story unfolded. The *Mass Effect* trilogy (2007-2012) revolutionized narrative-driven gaming by allowing players to make decisions that affected the overarching story, character relationships, and even the fate of entire civilizations. Choices made in one game carried over into subsequent titles, creating a sense of ownership for the evolving narrative. Open-world games also increasingly incorporated story elements. *Red Dead Redemption 2* (2018) allowed player actions to influence the main character's morality, as well as the game world, based on how they treat others.

As interactive storytelling evolved, it attracted scholars from a wide range of fields such as literature, psychology, computer science, and game design exploring how to craft engaging, meaningful stories on interactive platforms. Often considered one of the pioneers of interactive storytelling, Dr. Brenda Laurel explored the application of Aristotle's dramatic principles to human-computer interaction, particularly in her 1986 dissertation "Toward the Design of a Computer-Based Interactive Fantasy System" and her 1991 book *Computers as Theatre*. In it, Laurel applies Aristotle's six elements of drama to interface design, arguing that they map onto components like user actions (plot) and system responses (character) to create engaging and meaningful user experiences. She further parallels Aristotle's Four Causes framework to interface design to illustrate how technology and design serve purposeful, user-centered experiences. Dr. Janet Murray's influential 1997 book, *Hamlet on the Holodeck: The Future of Narrative in Cyberspace*, directly references Aristotle's concepts of dramatic construction in her discussion of the potential for interactive narratives. She uses *Poetics* as a foundational framework to discuss how digital environments can deliver structured and meaningful experiences while allowing for player agency. Yet, achieving narrative depth and complexity in interactive media while still allowing for significant player choice and agency is identified as a central challenge.[8] That same year, researcher Espen Aarseth published *Cybertext: Perspectives on Ergodic Literature* exploring how interactive fiction challenges traditional narrative structures. Aarseth points out that while interactive narratives offer new forms of engagement, they often struggle to provide the same depth and coherence as linear narratives.[9] As emergent gameplay—where players interact creatively with a game's systems to generate unexpected experiences and solutions—gained traction, game design shifted away from preordained stories to prioritize player agency, creativity, and the unpredictable interplay of systems. The sense of loss, triumph, frustration, or even humor a player feels in this approach tends to arise organically from unique interactions rather than as the result of a narrative arc designed to elicit specific emotional responses, as would be found in the structured, purpose-driven storytelling of Aristotelian drama.

A clear problem began to emerge: the more a guest is able to interact and affect the story world, the less coherent the narrative tends to be. As we see in classical stories with passive audiences, the flip side is also true: the more controlled the narrative, the less interactivity can be granted. Ruth Aylett and Sandy Louchart are credited with coining the term "narrative paradox" to describe the conflict between a pre-authored narrative and the freedom to interact with it.[10] The narrative paradox has since become a central consideration in game design, interactive fiction, and other forms of participatory storytelling where user choices intersect with a predefined narrative structure.

Thousands of research papers have cited "narrative paradox" or proposed models for overcoming the dilemma. Drawing from the work of Laurel and Murray, researcher Michael Mateas published "A Neo-Aristotelian Theory of Interactive Drama," in which he modified Aristotle's dramatic theory to address the interactivity added by player agency.[11] Clara Fernández-Vara's paper "Game Spaces Speak Volumes: Indexical Storytelling" proposes to bridge the gap between stories and games using spatial design.[12] For the field of interactive documentaries, David Millard suggested a model where narratives are viewed as interwoven threads, allowing for adaptability while maintaining coherence.[13] In their 2018 paper "Ludonarrative Hermeneutics: A Way Out and the Narrative Paradox," media psychologist Christian Roth and his colleagues analyze a cooperative game, exploring how its design navigates the tension between player freedom and narrative structure.[14]

Creators of every form of commercially available interactive digital narrative—from computer and video games to interactive films, virtual reality experiences, and story-driven apps—have faced the challenge of balancing player agency with an engaging story. In 2005, Mateas worked with Andrew Stern on a pioneering interactive drama called *Façade*. In the experience, the player assumed the role of someone who becomes entangled in their married friends' arguments. The story dynamically unfolded based on player interactions. Praised for its innovative methodology and its exploration of AI-driven character interactions, it also received criticism for its limited, predictable story outcomes that were essentially the result of following a branching narrative structure. In Netflix's *Black Mirror: Bandersnatch* (2018), viewers selecting between scenes or actions that determined how the story progressed found the branching choices sometimes led to confusing or unsatisfying conclusions, diminishing their engagement and emotional investment.[15] Live theatrical VR experiences like *The Under Presents* (Tender Claws, 2019) immersed guests in real-time stories presented in theatrical environments, but the unpredictability of the guest interactions led to fragmented and less impactful storytelling.[16]

The narrative paradox even extends to live theater where the emerging medium of immersive theater productions allow audiences to navigate a physical space as part of the production. Often cited as a defining example of the genre, Punchdrunk's *Sleep No More* sent masked audiences roaming freely through intricately designed environments inspired by Shakespeare's *Macbeth*. Like an open-world game, it lacked a singular linear story. Theater company Third Rail Projects tackled the challenge of spatial narrative slightly differently. In their immersive production *Then She Fell*, groups of participants started at different locations in the venue and then followed their specific routes in the overlapping narrative arc. However, the need for synchronized, multi-entry stories flattened the traditional emotional arc. Both productions limited actor dialogue, making them more thematic spectacles than conventional narratives. Similar to emergent gameplay, audiences find meaning in these experiences through their interpretation of the events they encounter.

Some scholars question if traditional story techniques are needed to evoke emotions as powerful as linear narratives. In her book *How Games Move Us*, Katherine Isbister explains how game developers rely on choice and flow to enhance players' emotional experiences. Others have argued that traditional story techniques may not allow meaningful storytelling with high interactivity, since in traditional narratives meaning depends on carefully constructed arcs and these arcs tend to disintegrate when a significant amount of agency is given to the player or reader. A key perspective on this issue comes from the field of ludology (game studies), where some theorists assert that the structure of a story, which requires a planned sequence of events, is fundamentally at odds with player freedom, which seeks to diverge from predetermined paths and offers emotional satisfaction through that sense of agency. Scholars like Jesper Juul have explored the limits of storytelling in interactive media, suggesting that complex narratives struggle to thrive under conditions where player actions must be accommodated.

Academic research from disciplines of narratology, ludology, human-computer interaction, media studies, and others tackling the narrative paradox have documented and expressed similar challenges. Across the majority of these diverse investigations, there is one common trend: each of these efforts remains firmly fixed on the assumption of primacy on plot and character.

And so the question remains: can guests meaningfully influence the progression of a story in such a way that the narrative remains engaging and coherent? Dr. Janet Murray envisioned a future where narratives can adapt fluidly to player inputs. Literary scholar Maria-Laure Ryan's vision is rather different and suggested a more problematic future for electronic narrative: "The aesthetic criteria of interactive drama will not be those of classical drama; the future of the genre will be as a game to be played and an action to be lived, not as a spectacle to be watched." But, she asks, "will this involvement be a source of aesthetic pleasure—will the game, in other words, be worth playing at all?"[17] Yet, stories were once inseparable from our daily lives.

SCIENCE AND STORY

In preliterate societies, narrative and natural phenomena were deeply intertwined. Myths did more than entertain; they were ways communities understood and explained the world. Stories blended what we now separate as "story" and "science." A myth about a thunder god, for example, provided an explanation for natural events like storms while also conveying moral or social lessons. The Babylonian creation story depicted the god Marduk defeating the sea goddess Tiamat to achieve cosmic order. In ancient Egypt, the annual flooding of the Nile was seen as the tears of Isis mourning Osiris, turning a crucial agricultural cycle into a narrative of loss and renewal. For sailors, stars were vital navigation tools, and the myths surrounding them provided a sense of place on the open seas. Story and understanding were inseparable and interactive; to recount a myth was to engage with the forces shaping daily life.

In ancient societies, natural occurrences were not just observed; they were immediately integrated into a lived narrative that gave them meaning and purpose. On May 28, 585 BCE, the Lydians were in the midst of a years-long battle with their neighbors, the Medes, when a solar eclipse transformed the skies over what is now known as the country of Turkey. As day suddenly turned into night, the warring parties saw it as a powerful, divine message, interpreting the sudden darkness as an omen that called for peace.

Just as the Lydians and Medes "played out" the implications of the eclipse within their cultural framework, participants in interactive drama are drawn into a story where choices and interpretations have tangible consequences. Ryan's "game worth playing" question of interactive drama recalls this ancient paradigm: stories are meant to be lived and acted upon, not simply watched. In both cases, whether in the ancient eclipse or in modern interactive narrative, the value lies not in passive observation but in active involvement and meaningful engagement.

The assumption of linear, centralized storytelling as the primary model began to solidify with the formalization of narrative principles in ancient Greece, particularly through Aristotle's *Poetics*, and became more entrenched as written traditions developed, especially in Europe. The rise of the novel in the 18th century further cemented the idea that a story should revolve around a central character with a continuous, linear plot, as novels focused on detailed character development and plot progression. This linear model grew in dominance, particularly in the West, as print culture emphasized coherence and sequence—characteristics that a written, linear narrative could easily support.

AN ARISTOTELIAN LIMP

The contrasting legacies of Aristotle's contributions to both science and drama since the 4th century underscore an important dynamic. Though originating from a shared foundation of systematic inquiry and observation, these two bodies of work have diverged markedly in their trajectories.

Aristotle's theories in his treatise titled *Physics* were repeatedly tested and ultimately reshaped by new empirical discoveries, leading to revolutionary paradigms. Concepts like the Four Causes lost prominence as thinkers like Galileo Galilei and René Descartes began to question Aristotelian physics, especially its reliance on purpose. Descartes introduced a mechanistic view of nature, emphasizing matter and motion governed by mathematical laws rather than inherent purposes. Later, Johannes Kepler formulated mathematically based laws of planetary motion, which calculated orbits as elliptical. This undermined the earlier idea that the heavens were designed with perfect purposes and forms and therefore must be circular, the supposed perfect shape. Shifts like these set the stage for the complete paradigm change brought by Newton, whose three laws of motion redefined motion and force in purely mathematical terms, without invoking purpose or inherent nature at all. Objects no longer moved to fulfill a final cause but behaved according to consistent, observable laws that applied uniformly across the cosmos.

Mathematics, the universal language of abstraction, evolved alongside physics. Early tally marks served as placeholders for concepts, enabling us to communicate and simplify complex ideas with symbols. In Aristotle's time, equations were reflections of static relationships. His contemporary, Euclid, developed a systematic approach to geometry, describing relationships through diagrams. For instance, the Pythagorean Theorem was demonstrated through geometric proofs rather than written as $a^2 + b^2 = c^2$. Aristotle's later dramatic theories mirrored this approach, providing a structured foundation for storytelling that emphasized plot, character, and catharsis as fixed elements designed to evoke specific emotions. As mathematics advanced, equations evolved to

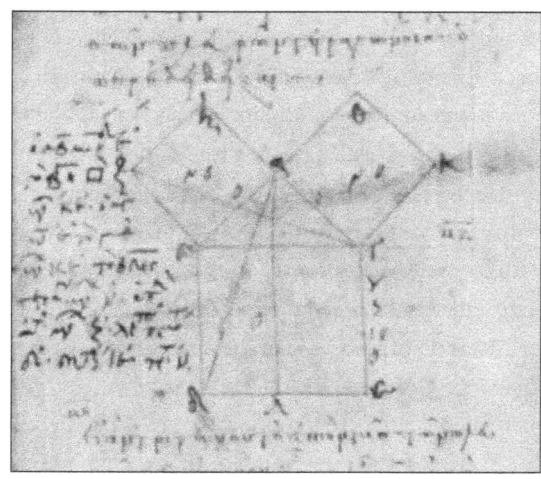

→ FIGURE 5. Excerpt from Proposition 47 of Book I of Euclid's *Elements*, demonstrating the Pythagorean Theorem (Euclid, 300 BCE).[18]

represent an understanding of dynamic relationships. Newton's second law of motion, $f=ma$, describes how forces cause changes in motion over time, emphasizing the continuous interplay between force (f), mass (m), and acceleration (a) rather than a fixed state. The emergence of algorithms extended these principles, combining equations into procedural methods capable of driving technological advancements and solving increasingly complex problems. Einstein's famous equation $E=mc^2$ further pushed this evolution, showcasing a reversible relationship between mass (m) and energy (E) that was inconceivable within earlier frameworks. Complex systems theory builds upon the scientific and mathematical advances of Einstein's time, focusing on the behavior of interconnected and dynamic systems—both natural and social—where outcomes are often unpredictable due to nonlinear interactions. In the context of human societies, researchers in the area of complex systems analyze the intricacies of human behaviors as collective properties of those systems.

In stark contrast, Aristotle's *Poetics* has endured as a seminal text in the study of literature and theater in the West. Storytellers have clung with reverence to the dramatic theories the work set forth, treating it as a near-sacred text that continues to guide narrative structures and dramatic theories with little challenge over the millennia. The reason for its lasting appeal in dramatic narratives seems to lie in the types of truths sought and how they were pursued. This may also explain their replacement in the sciences, as the scientific method with its values for empirical and experimental replicability became the industry standard. As tools, methodologies, and mathematics advanced, more accurate models replaced Aristotle's somewhat speculative approach, but the arts often continue to eschew quantifiable models in favor of subjective interpretations in regards to human conditions and emotions. Thus, his insights still resonate because they capture enduring aspects of human nature.

The uneven scrutiny of Aristotle's theories—vigorous in the sciences yet deferential in the arts—has led to a peculiar asymmetry in intellectual progress. It's as if humanity's pursuit of knowledge limps, advancing robustly on one leg while the other steps more cautiously, bound by the gilded chains of classical thought. Scientists have ventured into a challenging domain where established truths, once endorsed by eminent scholars, have been upended. Yet, the vast majority of interactive storytellers cling to ancient methodologies for crafting narrative, citing the unpredictable nature of humanity as a rationale for resisting change, even in the face of the transformative impact that today's technology can—and will—have on our stories. This challenge has finally caught the global attention of researchers, philosophers, and storytellers, yet even they seem unprepared to challenge the importance Aristotle placed on plot and character.

EARLY NARRATIVE SYSTEMS

Early complex systems balancing player agency with a coherent story did not involve computers at all, but rather live interaction within gameplay settings. The tabletop Role-Playing Game (TTRPG) *Dungeons & Dragons*, first published in 1974, relies on a Dungeon Master (DM) who adapts the story in response to players' actions. The DM weaves unexpected player choices, such as exploring side quests or inventing new objectives, into the overarching narrative. Live Action Role-Playing (LARP) is a different form of role-play that focuses heavily on physical immersion and collaborative storytelling. Unlike TTRPGs, LARPs don't typically have a central DM who controls the story moment-to-moment. Instead, players embody their characters in a shared space, interacting directly with one another to shape the narrative through spontaneous, in-character decisions. Nordic LARP in particular emphasizes deep emotional engagement and co-creation of the story. It also requires players to manage the narrative balance collectively. There is less emphasis on structured objectives and more on exploring broader character relationships and thematic experiences. These narrative systems involve multiple interconnected components that interact in dynamic and often unpredictable ways. In both forms, the human element plays a crucial role in evolving the story experience naturally based on player interaction.

Researchers of early computational interactive narrative systems created a similar role of "drama managers" within their project to keep the story engaging and meaningful, albeit through algorithmic means rather than human intuition. Dr. Joseph Bates' influential research initiative the Oz Project at Carnegie Mellon University in the 1980s and 1990s was dedicated to creating "believable agents," a term coined by Bates to describe computer generated characters designed to respond believably to user input. The project aimed to advance the fields of artificial intelligence (AI) and interactive storytelling by exploring

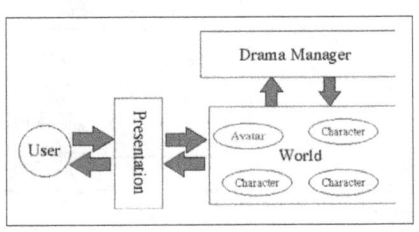

FIGURE 6. High-level diagram of the Oz Project interactive architecture wherein the Drama Manager changes the simulated World that is presented to the user (Mateas, 1997).[19]

how AI-driven characters could behave in ways that felt emotionally engaging and natural within a story. The Oz Project's research contributed to the development of virtual characters and the necessary systems to support them, which was groundbreaking for interactive media at the time.

While Bates focused on character development, Michael Mateas and Andrew Stern explored interactive plots. In their 2000 paper, *"Towards Integrating Plot and Character for Interactive Drama,"* Mateas and Stern focused on combining natural language processing (NLP) with AI-driven story structures. This approach allowed characters to respond dynamically to player input while advancing a cohesive narrative arc, bridging the gap between emergent character behavior and predefined plot points. Their work laid the conceptual groundwork for the previously mentioned interactive drama, *Façade*. By intertwining narrative progression with

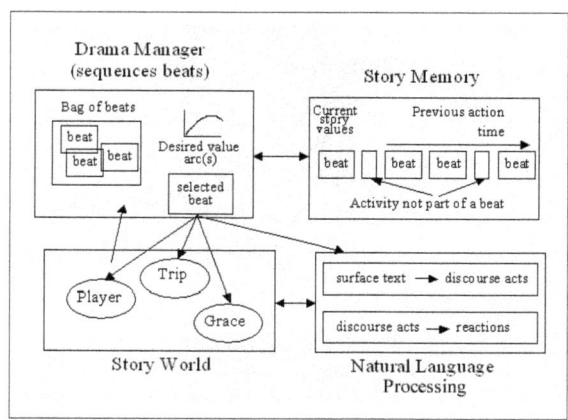

→ FIGURE 7. Major components of the *Façade* interactive drama architecture (Mateas and Stern, 2002).[20]

real-time character interaction, Mateas and Stern pushed the boundaries of player agency and emotional engagement.

The research of Dr. Mark Riedl, professor in Georgia Tech's School of Interactive Computing, has focused on the area of systemic story generation, often exploring how AI can generate coherent narratives autonomously or in collaboration with human users. His interactive storytelling system *Scheherazade* learns from crowdsourced data to produce coherent, logical narratives.[21] By processing multiple examples of a given type of story, the system moved beyond the rule-based systems or statistical models of the time to harness human narrative intuition on a large scale. Riedl and Young's paper "Narrative Planning: Balancing Plot and Character" proposed a narrative planning model that integrates classical

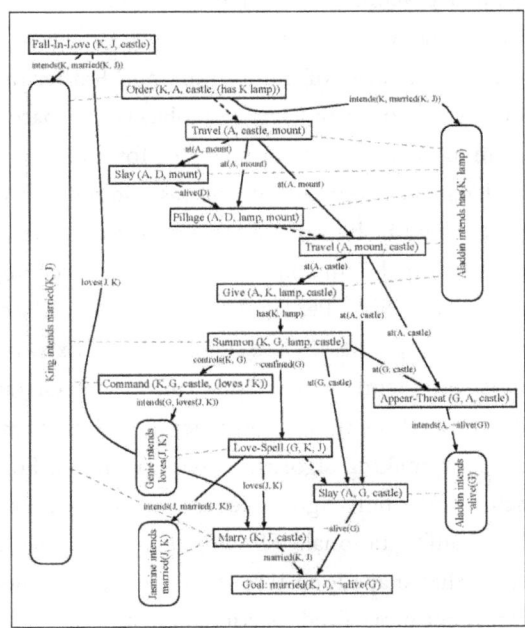

FIGURE 8. Fabula plan representation of the story used in the experimental Intent-based Partial Order Causal Link (IPOCL) search planning algorithm for narrative generation (Riedl and Young, 2010).[22]

planning techniques with character goals and actions to create more engaging and believable narratives.[23] Riedl's work continues to inform the fields of interactive narratives and computational narratology as they expand into increasingly complex systems.

Together, these works illustrate an evolution from agent-centric models to integrated narrative systems and scalable, data-driven approaches. Each step in this continuum contributed to refining the tools and methodologies needed to create increasingly sophisticated and emotionally engaging interactive story experiences, all while trying to balance the technical challenges of AI with the artistic demands of compelling narrative creation.

Other approaches to interactive storytelling have emphasized human psychological dynamics. In his book *On Interactive Storytelling*, game design pioneer Chris Crawford emphasizes "verb thinking" instead of "noun thinking." Rather than thinking about our world only in terms of objects in space, he suggests also thinking about them in terms of forces and events—noting that all languages contain both nouns and verbs. He relies heavily on psychology in his constructs, suggesting character behaviors and interactions be implemented by using psychological models such as the Big Five Personality Traits, which scores a person for traits of Openness, Conscientiousness, Extraversion, Agreeableness, and Neuroticism. Using this model, characters' score values determine how they might react in various situations to influence decision-making algorithms within the AI. For example, a character with high Agreeableness might be more cooperative and less confrontational. The Big Five describes stable personality traits that shape long-term tendencies, but other models can help represent more fluid reactions. For instance, the Pleasure-Arousal-Dominance (PAD) model captures transient emotional states that fluctuate in response to situations. Using numerical models for emotional states is generally considered a necessity for creating interactive dramas with traditional coding methods. In a game like *The Sims*, quantifying each character's personalities and emotions enables the game's AI to make decisions that simulate human behavior.

These are only a few of the multitudes of angles from which the narrative paradox has been tackled—it has been explored from the basis of narrative intelligence, memory recall, and symbolic plan recognition. Some decentralize the story; others design computational models motivated by neural networks. Some redefine the role of the player; others shift the role of the author. While these methods inform the future of interactive storytelling, none resolve the paradox.

Going back to the research previously mentioned, Bates' work enriched character interaction, but it didn't directly address how to ensure story coherence when players drastically alter the narrative. Across the collection of papers published from the Oz Project, there were consistent challenges in making virtual agents behave convincingly without sacrificing narrative coherence. The scope of *Façade* was tightly constrained, focusing on a single, short story. It faced challenges that were both external (believability and coherence of the experience to the players) and internal (technical complexities such as real-time processing, natural language understanding, and drama management). Scaling this approach to larger, more open-ended narratives remains a difficulty. While Riedl's procedurally generated stories can be coherent, they often also lack the emotional depth and character believability needed to make the story truly engaging, leaving the paradox unresolved.

The careful reader will recognize that many of these limitations trace back to a reliance on *Poetics* as a baseline—a framework that was never designed to accommodate nonlinear, player-driven, or emergent narratives that characterize interactive media. In Bates's Oz Project, the goal of creating believable agents prioritized realistic character interactions, but *Poetics* provided little guidance on how to balance dynamic character behavior with narrative coherence. Similarly, *Façade* attempted to manage the paradox by tightly limiting the narrative's scope. Mateas and Stern essentially created a closed loop of interaction—a short narrative space where deviations could be managed without breaking the core story. However, scaling this method to larger, open-world narratives exposed the fragility of relying on classical dramatic structure. Riedl's procedural systems, like *Scheherazade*, approached the scaling problem by generating narratives algorithmically for greater flexibility. Yet, the absence of direct emotional modeling reflects another limitation of *Poetics*: its emphasis on plot-driven resolution over the subtleties of psychological depth.

By adhering, even implicitly, to the principles of *Poetics*, researchers repeatedly confront the limits of applying a rigid, ancient framework to a fluid, dynamic medium. Ultimately, the paradox persists not because of a failure in technical execution, but because the theoretical models guiding these efforts are misaligned with the fundamental nature of interactive narratives. Resolving the narrative paradox may require a departure from *Poetics* entirely.

CLASSICAL CATASTROPHES

In the late 1800s and early 1900s, new technologies gave scientists access to the invisible world of atoms, the smallest unit into which matter can be divided. It turns out that at this scale, things don't behave as the fundamental laws of classical physics predicted. What they were observing and measuring are not things that most of us are going to encounter in any way, shape, or form. This is the world of the subatomic. To get a sense of how small an atom and its subatomic parts (protons, neutrons, and electrons) are, imagine a massive sports stadium. If the entire stadium represents the size of an atom, the nucleus (which contains protons and neutrons) would be roughly the size of a marble or a small pea placed at the center of the field. And the electrons would be almost invisible specks flying around the vast, empty space of the stadium. Atoms are 99.9999% empty space.

Two specific subatomic experiments shook the physics world: blackbody radiation and double slit apparatus. Oversimplifying it, the blackbody radiation experiment is about understanding why an object glows the way it does at different temperatures. Since a blackbody object isn't a real object but a theoretical concept, imagine it as a box with a small hole that is completely sealed from external light and heat. This box is considered a "blackbody" because it absorbs all incident light that enters through the hole and reflects none of it.

As we start heating this box, the interior walls begin to emit radiation. This radiation eventually reaches thermal equilibrium, meaning the amount of energy emitted by the walls is balanced by the amount absorbed. The radiation that escapes through the small hole can be observed—this represents the blackbody radiation. Initially, the radiation might appear as a dull red; as the temperature increases, it shifts to orange, then yellow, and eventually to a bright blue-white when extremely hot. Scientists noticed that the color at which an object glows is related to its temperature. But there was a problem: classical laws didn't accurately predict the colors of light (or radiation) the box would emit at high frequencies. This was such a major failure of classical theory that it was referred to as the "ultraviolet catastrophe."

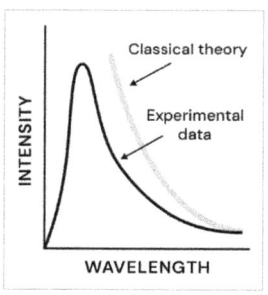

FIGURE 9. Diagram illustrating the difference between theoretical classical predictions and real-world experimental measurements of blackbody radiation.

The double-slit experiment likewise triggered an existential crisis in the scientific community by challenging their fundamental understanding of reality. To visualize this experiment, picture a wall with two narrow slits cut into it. Behind this wall is a second screen that catches whatever comes through the slits. When a beam of light is shined at the wall, light particles go through the slits. If light behaved only as particles, we would expect it to form two bright lines on the screen, one behind each slit—like if you threw sand through two gaps and saw two piles forming on the ground behind them. But instead, the light on the screen forms a lattice of light and dark stripes called an interference pattern. This pattern looks like what we see when waves overlap: the peaks of waves combine to make brighter areas, and where a peak and a trough meet, they cancel each other out, making darker areas. What was baffling was that when scientists measured which slit individual particles passed through, the interference pattern dis-

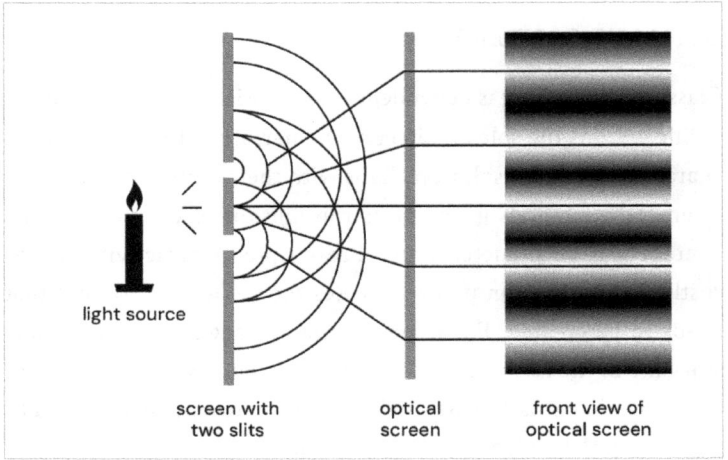

FIGURE 10. Diagram of the double-slit experiment, demonstrating the wave-particle duality of light as it passes through two slits, creating an interference pattern on the optical screen.

appeared, replaced by a pattern consistent with particle behavior. These findings revealed that light—and later electrons—could exhibit both wave and particle properties depending on whether or not they were observed, fundamentally altering our understanding of matter.

After centuries of having mathematical "recipes" for how the universe worked, from falling apples to orbiting planets, it was as if the universe had a secret set of rules for the small stuff. Physicists were forced to rewrite their recipe book for the universe, leading to a new branch in the natural sciences called quantum physics.

In ways equivalent to the physics catastrophes, new technologies and techniques are introducing insurmountable challenges to classical dramatic theory: dynamic content generation, spatial narratives, and interactivity.

DYNAMIC CONTENT

Classical storytelling is dependent on authorial control. The storytelling team is the sole architect of the narrative. Every plot twist, character arc, and resolution aligns with their unified vision. Early interaction methods gave guests the illusion of control, but the content was all predetermined. Pre-scripted interactivity can be costly. Every decision point increases the amount of content that needs to be created. For example, the average length of a single storyline in the *Choose Your Own Adventure* book such as "The Case of the Silk King" is 19 pages, whereas the entire publication is 115 pages long.[24] In video games, a standard script can range from 100-200 pages, but a branching narrative game like *Fahrenheit* (2005) required a 2,000-page script with added production costs for voiceovers and motion capture. With advancements in digital technology, these complexities can now be expanded upon at a lower cost. Netflix's interactive episode *Black Mirror: Bandersnatch* boasted over a trillion different permutations that all led to five main endings at only twice the production costs and time.[25] They achieved this, in part, by building custom software called Branch Manager that both created the flow of pre-scripted events and streamlined the production process.

Digital tools not only reduce production costs but also open up new creative possibilities. Dynamic content generation refers to content that adapts and evolves in real time, responding to user input or external variables in ways that have not been pre-determined. Instead, the content changes based on the audience's choices. From procedural generation to AI-driven systems, this shift allows for unique, context-sensitive experiences that are not only personalized and immersive but also fundamentally shaped by the audience's participation.

Procedural generation refers to the use of algorithms to automatically create game content, such as worlds, levels, characters, or items, based on predefined rules or parameters. Early examples like *Rogue* (1980) generated random dungeons, monster encounters, and treasures to enhance replayability. By the 1990s, games like *Diablo* enhanced this with randomized loot and level layouts. As computational power grew, titles like *Minecraft* (2009) and *No Man's Sky* (2016) took it further, generating entire worlds, ecosystems, and weather patterns. When a player starts a new game in *Minecraft*'s survival mode, the game generates a unique world using procedural generation algorithms; no two worlds are identical (unless using the same seed[26]). The landscape, biomes, cave systems, and even the placement of resources are generated spontaneously. Villages spawn with various configurations of buildings, villagers, and layout—all generated based on the surrounding environment and biome. *No Man's Sky* generated over 18 quintillion unique planets, each with its own terrain, flora, fauna, and weather conditions using fractal-based algorithms as a core component of its procedural generation system.

Today, procedural generation is not limited to environments. Characters, quests, and stories can also be dynamically generated based on real-time input, creating ever-evolving gaming experiences. In games such as Ludeon Studios' *RimWorld* (2018), the core storytelling comes from its AI storytellers that act as the player's "unseeable God."[27] AI-driven characters, environmental factors, and player decisions combine to create unforeseen events, such as a peaceful colony suddenly descending into chaos due to a chain reaction of seemingly minor incidents—a fire breaking out, resources running low, or characters reacting unpredictably to stress. Players must adapt to challenges that were not explicitly planned by the game designers. However, the game does not create new characters, events, or mechanics during gameplay. It strictly uses the existing pool of predesigned elements and rearranges them in novel ways. *RimWorld* is fundamentally procedurally generated. Its dynamic elements, particularly the AI storyteller, add layers of responsiveness and emergent storytelling. In this case, the creators build a framework of systems, rules, and interactions that allow the guest to shape their own narrative organically.

Unscripted live content like improv, dinner theater, and LARP shares similarities with *RimWorld* in how they both create narratives through interaction within structured systems. Improv acting's seemingly instinctual content often uses established formats, prompts, or character archetypes that are drawn out by audience interactions. Attendees at dinner theaters are invited to solve the crime, usually with a predefined plot structure and performers improvising within those boundaries based on guest response. In LARPs, the world, rules, and roles are often preset, while the interactions and outcomes are generated by the players' real-time decisions, creating a structured form of spontaneity. Human-driven storytelling allows for a wide range of creativity and emotional nuance, while games rely on procedural algorithms and AI to simulate those dynamics. Both offer emergent storytelling but in distinct ways: one through human ingenuity and the other through system-driven simulation.

Fully dynamic content generation is largely theoretical at this point. This kind of system would generate entirely new content on the fly without relying on any prewritten material. Maintaining a structured, engaging experience remains a challenge because dynamic generation lacks the intentionality of predefined rules and assets.

Dynamic content reveals how systems that generate information in real time face challenges in maintaining narrative coherence, as traditionally understood and developed. The fluidity of dynamic content disrupts classical notions of story relying on carefully crafted plots.

SPATIAL NARRATIVES

Ingrained in our experience of theater is the idea that we are there to "see"; the original Greek word *theatron* literally means "seeing place." The limited visual area of a proscenium stage gives the viewer a focal point and the creators composition parameters. Expanding stories beyond the screen or stage and into the physical world, or into a virtual world that is spatially navigated, disrupts our expected experience of story as well as its structure.

The desire to expand beyond the confines of the red curtain predates virtual reality and digital technology. The Italian Futurists experimented with the audience-performer relationship starting in the 1910s, often involving the audience directly in performances or staging actions in public spaces to blur the lines between art and life. Their radical ideas laid a conceptual groundwork that influenced later avant-garde and experimental movements. The Living Theater, founded by Julian Beck and Judith Malina in 1947, explored the use of space and audience interaction well before the formal establishment of Environmental Theater as a recognized movement in the 1960s, led by figures like Richard Schechner. Site-specific theater evolved in the decades following, intentionally anchoring its narrative to a particular location to allow the space's history, architecture, and ambiance to interplay with the theatrical narrative. Anne Hamburger, the founder of En Garde Arts, was instrumental in pioneering this genre in New York City during the 1980s and 1990s. Hamburger's productions brought forth plays that not only utilized the city's diverse urban landscapes but also tackled pressing social issues of the time. Her piece *J.P. Morgan Saves the Nation* took place on the steps of the Federal Hall National Memorial in New York City, making both a historical and contemporary statement.

Early forms of spatial theater explored non-traditional spaces and audience interactions without sacrificing linear progression. By creatively integrating the environment and using space to enhance rather than disrupt the narrative flow, they set the stage for even more radical departures in theatrical presentation, leading to our modern immersive theater productions where audiences physically navigate environments, interact with actors, and make choices.

Consumer-viable virtual reality (VR), augmented reality (AR), and mixed reality (MR) has likewise expanded digital content into embodied spatial environments. These formats eliminate the fixed vantage points of traditional screens and stages, inviting players to inhabit stories rather than merely observe them. In VR games like *Beat Saber*, players wield virtual lightsabers to slice through blocks in rhythm with music, requiring full-body movement as they dodge, swing, and interact within a three-dimensional space. Social VR experiences such as *VRChat* allow users to explore virtual concerts, attend meetups, or visit user-created environments. AR games like *Pokémon Go* overlay digital creatures and elements onto the real world via smartphone cameras, generating events specific to a player's location. Unlike traditional media, spatial content breaks the boundaries of the page, screen, or stage. This shift transforms stories from a confined medium into an experience that merges with lived space, making the boundaries between reality and narrative nearly invisible.

Increasingly, audiences also navigate stories across platforms, media, and locations. Distributed narratives are very much a reflection of our contemporary media landscape, and they offer unique opportunities for immersive and participatory storytelling while also presenting distinct challenges in narrative deployment. In them, audiences engage with different parts of the story by moving through diverse spaces, whether they are digital environments, physical locations, or alternate formats. The story of Harry Potter extends from books and films to theme parks and interactive experiences. *Halo*, originally a computer and console game, expanded into a sprawling franchise that now includes comics, books, TV series, and Alternate Reality Games (ARGs). Storylines in the Marvel Cinematic Universe span films, streaming shows, comics, games, and even live experiences. For example, Wanda Maximoff's story begins in the film *Avengers: Age of Ultron*, deepens through subsequent films, and reaches emotional heights in the television series *WandaVision*. Meanwhile, Marvel comics explore her backstory, which the screen adaptations only hint at. And in the *Marvel Future Revolution* mobile game, Wanda appears as a playable character, offering fans a chance to directly shape her battles. At live events like Avengers Campus in Disneyland, parkgoers can interact with Wanda through scripted encounters that make them part of her story. Traversing these platforms, audiences gain a multifaceted understanding of Wanda's journey. These kinds of cross-platform stories transform a single character's arc into an expansive, participatory universe.

Attempts to integrate coherent narrative into spatial productions have included strategies such as bounded choices (when audiences can make choices within specific boundaries), overlapping narratives (where multiple storylines or scenes might occur simultaneously), and reactive characters (actors trained to adapt and improvise based on audience interactions). For generations of people accustomed to passively watching framed content, interacting with actors and spaces may not be a familiar or even desirable experience. Thrust into unfamiliar territory, they may struggle with the ambiguity of interaction "rules," the implicit and explicit guidelines that govern audience interaction in spatial experiences. These rules help audiences understand their role, how they should engage with the story, and the boundaries of their influence. Because traditional storytelling typically places audiences in a passive role, spatial productions must address the ambiguity of these new interaction dynamics. Spatial productions that fail to clarify these interaction rules risk leaving audiences confused.

Spatially navigating a story within or across environments introduces physiological challenges that compete for our attention and detract from story engagement. The concept of cognitive load, developed by Australian educational psychologist John Sweller, is based on the limitations of working memory—a core component of the brain's cognitive functions. When too much information is processed simultaneously, working memory can become overloaded and the ability to process emotions and stay engaged diminishes. In the case of spatial narratives, requiring the audience to make decisions on where they should move or if they should interact increases their cognitive load. Their focus then shifts from the story's emotional beats to logistical or navigational concerns. The lack of dialogue in immersive theater productions *Sleep No More* and *Then She Fell* had the unintended benefit of reducing the cognitive load of guests who were moving through the environment.

The spatial catastrophe arises when audience agency disrupts the narrative arc. Audiences may linger in certain areas, bypass key plot points, or engage with side elements that were never intended to carry the weight of the main story. Granting participants the freedom to move often undermines the coherence essential for emotional and intellectual engagement.

THE INTERACTIVE SURGE

Today, there is exponentially more digital content than physical content, and it is largely accessible at almost any moment. From touch and gesture controls on mobile devices to geolocated augmented reality experiences, interactivity dominates digital content. Once confined to simple clicks, people can now deeply personalize content, engage in live streaming with real-time feedback, and participate in multi-platform experiences where actions taken on one platform affect outcomes on other platforms and in other locations in real time. Both dynamic content generation and spatial navigation are rooted in interaction.

While traditional storytelling has struggled to incorporate interactivity, it has always been fundamental to gaming. Early computer games started with rudimentary interactions, often developed as demonstrations of computing capabilities. The iconic 1972 video game *Pong* allowed two players to interact simultaneously by controlling paddles to play a virtual game of table tennis. Using a simple black and white screen with a dotted line through the middle, *Pong* was revolutionary at a time when computers were primarily used for basic calculations and data processing. It showcased things we take for granted in the real world: real-time interactions, collision detection, and simple physics.

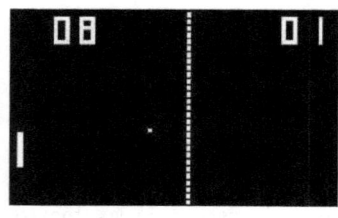

FIGURE 11. Screenshot of Atari's 2 KB game *Pong* (Coterhals, 2023).[28]

A decade later, strategy games such as *M.U.L.E.* allowed players to choose whether they would cooperate or compete as they managed resources on an alien planet. While graphics were still primitive by today's standards, the advancements in processing power in the 1980s prompted popularity in text-based adventure games, also known as interactive fiction, where players used simple two-word commands to influence the events of a branching

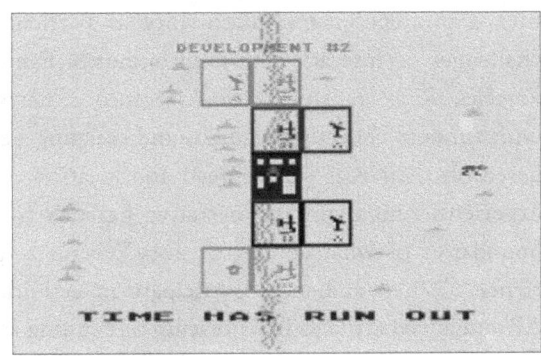

→ FIGURE 12. Screenshot of *M.U.L.E.* on Atari 8-bit PAL (Wikipedia contributors, n.d.).[29]

narrative story. As more visually engaging experiences began to be supported, commercial gaming shifted to focus increasingly on realistic graphics and complex gameplay. By the mid-90s, the differences between game and story narrative-driven works like *Myst* were sparking debate over the role of narrative in gaming.

As games evolved from the 1990s to today, interactivity expanded far beyond basic commands and simple branching paths to incorporate complex narratives, decision-making, and immersive environments with varying degrees of player agency. Computer programs, websites, video games, social media platforms, and mobile apps can all invite the user, player, or audience to participate in the stories they present. Every day, billions of people interact with content through these platforms, shaping and engaging with stories in unprecedented ways.

More and more, interactivity is no longer confined to gaming; it's weaving into unexpected places, reshaping how stories are shared and experienced. Museums now incorporate interactive storytelling through AR-enhanced exhibits where visitors scan artifacts with their phones to unlock historical reenactments or character-driven narratives. Experiential marketing campaigns turn a product into a micro-story as participants compete in challenges or interact with environments. Even the fitness app *Zombies, Run!* transforms exercise into a narrative adventure, with runners collecting supplies and escaping hordes of zombies based on their real-world speed and location. These interactive experiences highlight the narrative paradox by challenging the boundaries of what defines a "story" versus a "game" or "experience." When audiences participate in activities like scanning AR-enhanced artifacts in a museum or running from zombies in a fitness app, their actions are integrated into the experience. However, the story's coherence often depends on the creators' ability to predefine how these interactions fit into the narrative framework.

For example, in *Zombies, Run!*, the runner's pace influences the gameplay, but the "story" itself—delivered through audio prompts—is static and unfolds regardless of player decisions. The game mechanics (running, collecting supplies) give participants agency, but they rarely alter the narrative's trajectory. This blurs the line: is the user engaging with a story, or simply playing a game with a narrative backdrop? Similarly, a Coca-Cola vending machine that unlocks a drink through playful interaction might seem like a "story" moment, but without a meaningful narrative arc, it becomes more of a gamified experience than a true story.

Technology has not only expanded interaction in storytelling but has also integrated it into our everyday lives. Sensory feedback hardware like haptic gloves enables users to physically "feel" virtual environments. Beyond VR, haptics show up in everyday devices, like the subtle buzz of your smartwatch that reminds you to turn during navigation, creating a tactile interaction with digital content. IoT (Internet of Things) devices that transmit data over the internet without human interaction enhance this further—smart refrigerators suggest recipes based on the items you scan, while home assistants like Alexa or Google Nest become the narrators of your daily life, seamlessly blending digital prompts with physical actions. These technologies transform interaction from novelty into a seamless part of daily experiences, where storytelling opportunities exist at every touchpoint.

Humans are natural storytellers, and we layer meaning onto experiences even when no story is intended. People assign intentionality to actions or outcomes, even in ambiguous situations. Fans create personal interpretations of characters' arcs by connecting dots between films, TV shows, and comics. People document ordinary events on Instagram or TikTok; cooking dinner or walking a dog become micro-stories when imbued with meaning from captions, music, and filters. Whether through IoT devices, cross-platform universes, or everyday technology, humans instinctively create narratives by layering or extracting meaning. These interpretations transform functional interactions into emotionally resonant experiences, bridging the gap between utility and story. All this interactivity creates complexities that the *Poetics* can't account for any more than the Pythagorean Theorem, with its static geometric relationships, can describe the nonlinear interactions of components in complex systems.

Interactivity lies at the heart of the narrative paradox. It empowers audiences to shape the story, but this same freedom threatens coherence and diminishes its emotional impact. As narrative control expands from the hands of authors and out across systems, spaces, and audiences, the future of interactive storytelling depends not on refining classical methods, but on understanding and embracing the distinct nature of these evolving domains.

BENDING LIGHT

At first glance, the double-slit experiment might appear to show light bending. The interference pattern created suggests light is weaving its way through space, bending like water rippling around obstacles. But when scientists explore deeper, they can see that something far more profound is happening. Light isn't bending in the way we understand it. It's flickering between states depending on how we observe it—as a particle or a wave. The "bending" is a mirage, a hint of the deeper truth: reality itself is not as fixed or linear as it seems. When we stop fixating on light as a collection of particles and deeply explore its behavior, the illusion of bending isn't a trick—it's a clue pointing to the wavelike nature that defines its essence. Could the narrative paradox itself be a similar clue—that stories, like light, might not behave by the same Aristotelian rules once interactivity is introduced?

One of the most explicit and theoretical challenges to Aristotelian dominance in interactive narratives has come from Ian Bogost, a renowned scholar, game designer, and author. His 2007 book *Persuasive Games: The Expressive Power of Videogames* introduced the concept of procedural rhetoric. This approach highlights the unique power of games as a medium to simulate complex systems and allow players to explore ideas through interaction rather than passive observation. Bogost proposes that the rules and mechanics of a game embody values, ideologies, or perspectives. By interacting with the game's systems, players engage with those embedded messages. Shifting the focus from characters and plot to systems and rules begins to expand what storytelling can mean in interactive media.

The design of Lucas Pope's game *Papers, Please*, released in 2013, aligns closely with Bogost's principles of procedural rhetoric. In it, the player takes on the role of an immigration officer in a fictional dystopian country, balancing personal needs like keeping their family alive with the job's oppressive demands. It creates a deeply impactful story not through a fixed plot or character arc but through procedural storytelling, where the player's interactions with the game's mechanics and systems drive the narrative. *Papers, Please* received critical acclaim for its narrative innovation. Critics praised its ability to deliver a powerful story while maintaining player agency.

The success of *Papers, Please* raises the possibility of how systems whose mechanics are not driven by plot and character might represent the future of interactive narrative. Unlike traditional frameworks that rely on linear progression and emotional catharsis, procedural storytelling allows player-driven experiences to adapt to individual choices. Games like *Orwell* (2016) and *Not Tonight* (2018) also explored these systems, using mechanics to convey themes as effectively as traditional narrative techniques. With interactivity central to stories in digital media, non-Aristotelian systems offer a compelling alternative, creating stories that are not told but rather experienced. If dramatic stories can emerge from the dynamic interplay of player choices and systems without a guiding author, what deeper forces make them feel so inherently resonant and human?

Take a moment to question your assumptions about story. Consider the limitations that entrenched conventions impose on it. Are they bending the light? Is a radically different truth waiting to be discovered? For example, assuming that a story should revolve around a single protagonist's growth and journey limits the possibility of simultaneous, asynchronous development of multiple characters. Assuming that narratives must follow a linear progression restricts the opportunity to experiment with audience involvement in piecing together a story in their own order. Assuming that stories need a fixed beginning, middle, and end limits the potential content generated in response to audience choices. Assuming that an author should maintain control over the story constrains the role and functionality of the audience.

By challenging our assumptions about what has been successful in previous mediums, we open the door to evolution. Nowhere do we see this more than when we infuse uncertainty and unpredictability into the process of content creation. A willingness to depart from familiar methods and venture into unknown territory challenges the very principles that have guided authors and creators for generations. Yet, the history of storytelling is marked by moments when breaking the rules led to new forms and possibilities, leaving us to wonder: what might happen when we challenge the very foundations of narrative itself?

Think of the plot not as a preconstructed chain of events triggered by interaction. Instead, think of it as existing in potential, waiting to emerge based on the choices the player makes. A simple decision to seek revenge or to forgive can send the narrative down a new path—one completely unpredicted by the creator of the experience. Here, the plot isn't a map, but rather a field of potential outcomes. And so, as the player interacts, observes, and decides, the story unfolds not as a straight line, but as a series of possibilities actualized in real time.

Characters, too, take on new life in this space. No longer confined to the scripted arcs and destinies assigned by the author, they become responsive, evolving presences in the narrative. AI algorithms allow them to react, adapt, and remember what's been said and done. A character might remember a past slight or grow closer to the player based on shared experiences; they may alter their own motivations, driven not by the author's intent, but by the player's actions. Characters aren't merely acting out a predefined role but are like people in a world that remembers, building relationships that feel authentic and unique. They respond with depth, reflecting back the choices made, becoming mirrors of the player's decisions as much as actors within the story.

And the setting itself, traditionally a static background, is transformed into a reactive environment. It is no longer just the stage but a world that changes in response to its inhabitants. Mountains and cities, weather patterns and ambient sounds, all adapt to the guest's choices, perhaps to darken the sky or bring life to the landscape; maybe the forest is haunted by echoes of past actions. The setting is not fixed but rather a canvas that paints itself with each step, a co-creator in the narrative, one that holds memories of every interaction.

Dialogue, even, is freed from the constraints of prewritten lines. Conversations are fluid, informed by prior choices, allowing for exchanges that are truly unique. This isn't dialogue simply delivered; it's dialogue that spontaneously reacts to the moment and continues to evolve, that feels as though it could go in any direction. It's a conversation not just with a character but with the world itself, where each word matters, each pause holds weight, and each response could change the course of the story.

Freed from a linear format, stories transform, becoming a system of interconnected possibilities. In this dynamic form, a story isn't told, but is a world that's alive with potential where meaning emerges uniquely with each engagement. Stories, plot, character, setting, and dialogue are interdependent, existing for the guest to experience and shape.

In this space, we stand at the edge of an era that demands we rethink everything and embrace the unknown, just as scientific exploration has always done. Like light, stories are more than what they seem at first glance. To see their true nature, we must let go of old assumptions and embrace the waves of possibility that lie ahead.

QUANTUM NARRATIVES

Newtonian physics described a world where events and objects behave according to straightforward cause-and-effect relationships. Time was thought to move steadily, like a universal clock, and space was seen as a constant, unchanging grid through which objects moved. In 1905, Einstein introduced equations that showed how time and space could change depending on the speed at which one observer is moving relative to another. One of the most famous results of this theory is time dilation—the idea that as an object approaches the speed of light, time actually slows down for it relative to a stationary observer. This concept demonstrated that time intervals and distances weren't fixed; they varied according to speed.

Ten years later, in his General Theory of Relativity, Einstein extended these ideas to include gravity, fundamentally redefining it. Instead of seeing gravity as an invisible force pulling objects toward each other, he described it mathematically as the curvature of spacetime. This replaced the predictable grid of Newtonian physics with a flexible, interconnected universe, where time and space were no longer constants but variables shaped by motion and gravity.

Suddenly, everything depended on perspective, and events could look different depending on where you were and how fast you were moving. Einstein's theory didn't just add new details to physics; it redefined the fabric of reality—for scientists, anyway. For the rest of us without an understanding of advanced physics, the world didn't change; clocks still ticked, apples still fell, and our everyday experiences of time and space remained as they always had been.

DETERMINISTIC VS. PROBABILISTIC

A profound yet straightforward distinction allows the classical physics of Newton and the quantum mechanics of Einstein and others to coexist with minimal conflict: the scales at which these theories apply. Classical physics governs macroscopic phenomena while quantum mechanics deals with subatomic particles. In the classically visible world, the behavior of systems can be precisely predicted if the initial conditions are known. It is deterministic. However, quantum mechanics necessitates a probabilistic approach because the behavior of particles and waves at small scales can only be described in terms of probabilities, not definite values, until measured.

The distinction between classical and quantum domains not only expands our understanding of the universe but also provides a conceptual lens that could revolutionize how we approach interactive stories. Given the same starting point, the laws, elements, and mechanics of classical physics will always produce the same result. Classical stories are similar in that the outcomes are entirely predetermined, regardless of reactions. A theater production can also be predicted with certainty: the show starts, the audience sits and observes, the actors speak the words and move in the trajectory and the manner directed. The immediacy of the audience may have some effect—perhaps they don't laugh at a particular comedic moment—but for the purposes of systemic evaluation, the initial conditions of the system can be predicted. Classical stories, like classical physics, follow a deterministic model.

This orderly experience provides stability, much like the concepts of absolute time and space in the physical world. However, as we move away from stories confined to deterministic imitations of reality and into content that reacts to our presence, content becomes relative and probabilistic. A single interactive story can be experienced differently depending on what characters we follow, where we enter the story, or how we interpret its unfolding events.

Aristotle's concept of *mimesis*, Greek for imitation, reflects a view in which art and literature aim to approximate reality by capturing its essential patterns and truths. Through re-creation and interpretation, artists distill the complexities of the world into representations. However, just as mimesis is an interpretation of reality rather than a perfect reproduction of it, classical physics emerges as an approximation of quantum mechanics when applied to larger systems, smoothing out quantum uncertainties at observable scales. Rather than viewing Aristotle's framework as outdated, it can be seen as a foundation—a reflection of the best understanding of narrative at the time, perfectly suited for linear forms like theater, film, and literature. Quantum narratives, much like quantum mechanics, build upon this classical groundwork, allowing for stories that adapt and reflect the complexity of guest engagement and nonlinear progression.

The narrative paradox arises from the struggle to reconcile the deterministic traits of classical storytelling with the probabilistic nature of interactive narratives. Just as we use different rules for the macroscopic and subatomic worlds, we should explore doing the same for stories.

Recall how light can be interpreted in two different ways simultaneously. In the classical view, light behaves like a wave, explaining phenomena like reflection, refraction, and the colors of a rainbow. However, on a microscopic level, light also behaves as discrete particles governed by quantum mechanics, explaining phenomena like the photoelectric effect. Despite their differences, both descriptions are correct and coexist, providing a fuller understanding of light in different contexts.

In storytelling, society has extensively explored the visible classical elements such as plot, characters, and structure—which are akin to observing light as a wave. These elements are the straightforward, tangible aspects of a narrative that guide the audience through a predictable, linear experience. However, just as light also behaves as particles at a quantum level, narratives have the potential to behave differently in interactive environments. These deeper, quantum-like elements influence the narrative's direction in ways that aren't immediately obvious but are essential to its full impact. Both the predictable, classical components and the unpredictable, subtler forces can coexist.

Viewing narratives through this dual lens, we can embrace both the structured predictability of classical tales and the fluid, adaptive possibilities of quantum narratives. Rather than forcing classical canons onto mediums that follow fundamentally different rules, we can understand that they function differently within their respective domains.

BREAKING NEW GROUND

Progress often doesn't come from perfecting what we already know but from daring to question it and using the tools at hand in new ways. Einstein didn't need groundbreaking instruments to formulate his theories; he worked with the same mathematics and principles already available. At the time, no tools could even measure the effects he predicted. It took decades for technology to catch up, with inventions like atomic clocks and particle accelerators confirming his ideas with breathtaking precision. With stories, we have a similar opportunity: to move beyond established norms by reinterpreting traditional frameworks like character and plot through a new lens.

Unlike the precise measurements of physical properties like mass or velocity, human behavior involves factors difficult to quantify and assess objectively such as emotions and motivations. Philosophers of science such as Thomas Kuhn and Karl Popper highlight that the methodologies of "hard" sciences like physics and chemistry rely on empirical data and predictable outcomes, whereas the "soft" human sciences such as psychology must navigate the complexities of meaning, intention, and cultural context, requiring interpretive methodologies.

Psychology has long been considered the most applicable natural science for crafting dramatic narratives, offering insights into human behaviors. Writers use these psychological insights to shape plot and conflict. However, as we begin to think about stories as dynamic systems, the inward focus of psychology may not be the most useful tool. The tools may not exist yet, as in Einstein's case, but what else can we use?

Despite the universe's apparent complexity, mathematics has distilled its intricacies into elegantly simple formulas by noticing patterns, questioning their properties, and creating tools or theories to generalize and explain them. But the process of getting to that simplicity is complicated. Einstein's famous equation $E=mc^2$ emerged from a series of theoretical and mathematical challenges such as the need to reconcile the laws of motion with the constancy of the speed of light, and the need to reformulate the concept of mass-energy equivalence. As complex as the universe is, scientists currently believe that all matter in it is ultimately composed of the same fundamental constituents: four fundamental forces, twelve particles, and one hundred eighteen elements. We humans are no exception. The same forces and particles that explain the vastness of the cosmos also govern the cells, molecules, and atoms within the human body. Gravity, which keeps our feet grounded on Earth, also controls the motion of planets, stars, and galaxies, shaping the structure of the universe. Electrons, which are crucial for nerve signal transmission in humans, also participate in nuclear fusion reactions that power the stars and contribute to electromagnetic fields that influence cosmic phenomena such as auroras. The mathematical frameworks that simplify our understanding of the universe illuminate the interconnectedness of all matter.

The deep connection between mathematical theory and the natural world is manifested across natural phenomena like the Fibonacci sequence,[30] which appears in natural patterns like shell spirals and the arrangement of leaves, and fractals. Fractals, patterns that repeat infinitely at different scales, help model complex features such as cloud formations, mountain ranges, and coastlines. They provide a basis for understanding how these structures form and persist over time, despite being influenced by numerous seemingly chaotic environmental factors.

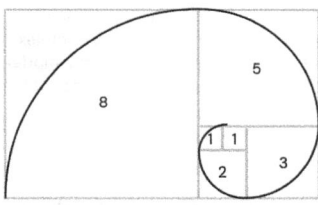

↑
FIGURE 13. Diagram of the golden spiral, illustrating the proportional relationships between consecutive Fibonacci numbers.

→
FIGURE 14. AI-generated geometric starburst pattern with fractal-like structures.[31]

The concept of similar patterns also extends to human behavior. Patterns can be observed in various aspects of human activities, relationships, and social structures. Network theory models human relationship patterns, revealing fractal-like properties. Social networks have nodes (people) and connections (relationships) that show self-similarity at different scales. The way people group into families, social circles, communities, and even larger societal structures can be understood through repeating patterns of how individuals connect and interact. Cognitive biases and heuristics that guide human decision-making, such as the tendency for conformity or fear response, tend to follow predictable patterns that can be statistically modeled and anticipated, even in the face of uncertainty.[32]

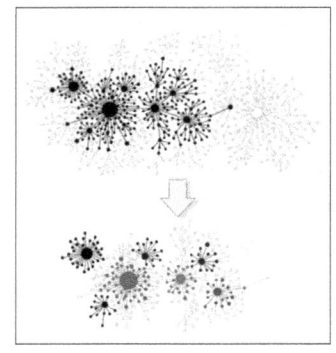

FIGURE 15. Schematic illustration of geometric self-similarity in complex networks, based on the fractal model of nested BA networks (Fronczak et al., 2024).[32]

Patterns in human behavior are not just coincidences and curiosities; they provide actionable insights. Behavioral economics, for example, uses these patterns to understand and predict economic decisions, such as the consistent ways in which people respond to incentives and risk. Seemingly irrational decisions often adhere to predictable biases, like the overemphasis on short-term rewards or aversion to loss. Game developers in particular leverage these patterns by designing experiences that nudge players towards specific choices. By framing gameplay options to emphasize potential losses rather than gains, they are able to predict and influence players' decision-making in subtle yet powerful ways.

In shifting from the predictable flow of linear, classical narratives into a space where control dissolves, interactivity in all forms introduces a myriad of possibilities. How do we channel that unpredictability into something emotionally resonant?

LAW OF ENGAGEMENT

Aristotle believed the purpose of tragedy was to evoke pity and fear in the audience. Given his belief that everything has an intrinsic purpose that defines its nature and value, we can hypothetically extend his views on purposeful design to other genres. We might surmise that he saw comedy as fulfilling a role in purging societal tensions through laughter, thereby promoting social harmony. Perhaps he thought the epic form, with its grand narratives and heroic figures, served to inspire and instill virtues such as bravery and wisdom.

A similar concept of dramatic purpose can be found in other parts of the world. In ancient India, *Natyashastra*, the treatise on the performing arts by Indian sage and scholar Bharata Muni articulates the concept of *rasa*, a term referring to the emotional states that a performance is intended to induce in the audience. There are eight primary *rasas*, each corresponding to a basic human feeling: love (*Shringara*), humor (*Hasya*), fury (*Raudra*), compassion (*Karuna*), disgust (*Bibhatsa*), horror (*Bhayanaka*), heroism (*Veera*), and wonder (*Adbhuta*). The audience, through experiencing these emotions, comes to a deeper understanding of the human experiences depicted on stage. A medieval Japanese Noh theatre master, Zeami Motokiyo, developed a framework that includes the concept of *yugen*, which suggests a profound and mysterious sense of beauty that evokes a deep emotional response.

While the expectation is that dramatic stories are designed to evoke emotions, authors across mediums have harnessed their craft to achieve a variety of intentions. Classical authors like George Orwell have used novels as tools for social critique, embedding warnings about political dangers in works like *1984*.

Filmmakers, particularly in documentaries, have used their craft to illuminate pressing social issues and inspire change. For example, Ava DuVernay's *13th* uses archival footage, interviews with scholars, activists, and politicians, and statistical analysis to inform and mobilize audiences. Cirque du Soleil shows aim to inspire awe and wonder through breathtaking acrobatics and innovative spectacles. Beyond entertainment, stories humanize brands by conveying their values through origin stories and mission statements. They make data or abstract ideas accessible and relatable. Public health campaigns use stories to motivate healthier choices like quitting smoking. Stories build persuasive legal cases, drive fundraising efforts, and make learning for students engaging and memorable. But in order to achieve those end goals, they must first engage.

Theorists and creators of stories have inadvertently aligned on a universal purpose: the primary goal of story has been to simply *engage the audience*, be it emotional, cognitive, behavioral—or a combination of these.

Whether a narrative follows traditional structures or breaks them, whether it evokes catharsis as in Aristotle's *Poetics*, induces rasa in the *Natyashastra*, or challenges the audience's critical perspective as in Brecht's alienation effect, the ultimate purpose is to hold and maintain the audience's attention. Engagement is a constant across diverse forms, achieved through different methods depending on cultural context, story structure, or even audience expectation.

Engagement often feels instinctive, but what truly pulls an audience in? Maybe it's the way certain moments resonate through words such as Hamlet's famous soliloquy on existence that begins "To be or not to be," or how the unfolding of events aligns with a particular rhythm like the blend of action with deep philosophical questioning in the iconic film *The Matrix*. Could it be the texture of the world, the subtle interactions between characters, or the atmosphere that sets the tone? The highest-grossing Broadway show of all time, *The Lion King*, is known for Julie Taymor's innovative staging and large-scale puppetry and the memorable music composed by Elton John and Hans Zimmer. But not all stories follow a Western or traditional structure, and they still manage to hold attention. Modern stories eschewing singular cause-and-effect plots have been replaced by intricate character arcs that evolve over multiple seasons, formats, and events. There's something more at work, a web of underlying, invisible mechanisms that sustain engagement, whether it's emotional, intellectual, or both.

THE INVISIBLE STORY

Scientists have been studying our engagement with stories for over a century, but breakthroughs in recent decades have accelerated the field. It turns out that stories serve as both a mirror and an embodiment of the human experience. Their dual role as reflection and structure is deeply tied to the way the brain processes them. What makes stories so powerful is their ability to align seamlessly with how we make sense of the world around us. In essence, stories are a neurological experience—universal, timeless, and, until recently, largely invisible.

From the moment a story begins, the brain seeks to find patterns and make sense of the events unfolding. Our brains have an intrinsic desire to process information in sequences, specifically the hippocampus, which organizes sensory input into logical sequences,[33] while the neocortex plays a crucial role in recognizing patterns, enabling us to interpret situations quickly and respond appropriately.[34] Familiar sequences activate event schemas—mental structures that allow us to predict what will happen next—enabling us to process information efficiently and creating a sense of satisfaction as a story unfolds. Patterns are why structures like Freytag's Pyramid or its modernized variations have been incredibly effective—not because plot points happen in a specific order, but because human brains seek familiar templates to guide our expectations, particularly when processing stories.[35] The predictability of traditional narrative arcs, with their familiar cadence of conflict and resolution, resonates with our evolutionary predispositions towards seeking patterns. Unconventional story structures, like collage narratives or fragmented stories, still keep the brain engaged by prompting it to fill in gaps or draw connections, activating our pattern-recognition faculties and keeping us actively involved in making sense of the information.

As a story builds tension and conflict arises, the brain responds by heightening focus in several ways. Conflict activates the prefrontal cortex, the area responsible for planning and decision-making.[36] This is why stories rich with obstacles and stakes captivate audiences and hold their attention. During a conflict, the brain experiences heightened emotional and physiological arousal. This is often accompanied by increased activity in the amygdala, which is responsible for processing emotions such as fear and excitement.[37] The anterior cingulate cortex monitors discrepancies between predicted and actual outcomes, rewarding the brain with dopamine spikes for resolving mysteries or plot twists. Neurotransmitters like dopamine function to drive curiosity, physiologically keeping audiences engaged. Dr. Paul J. Zak, who has focused decades of research on the neuroscience of narrative, found that a story with a dramatic arc caused an increase in both oxytocin and the stress hormone cortisol, which our brains produce during tense moments in a story to enhance our focus.[38]

When the audience bonds with a character, the brain activates mirror neurons, allowing us to experience their emotions as if they were our own. Even when it comes to stories without conflict or Western structure, the mirror neurons in the brain activate when we observe someone else performing an action, allowing us to empathize as if we are experiencing their actions ourselves.[39] Mirroring provides a neurological explanation for the psychological concept of narrative transportation—the phenomenon of becoming mentally and emotionally immersed in a story. Zak's earlier research identifying oxytocin as the "moral molecule" that enhances trust and social cuing offers a complementary neurochemical perspective, suggesting these mechanisms may work together to foster genuine connections with fictional characters and inspire real-world changes in behavior.[40]

At the emotional and narrative peak, the engagement of multiple brain regions also peaks. The amygdala intensifies emotional responses, while the brain's reward system prepares for catharsis. Cortisol spikes during the climax but recedes during the resolution, replaced by dopamine (reward) and oxytocin (connection). Reaching catharsis is deeply connected to the brain's emotional processing centers and, here, the medium also plays a role. By providing a safe and controlled space, individuals are able to work through complex feelings. Controlling the environment is central to various therapeutic modalities, such as Emotion-Focused Therapy, which aims to help clients become aware of and express their emotions in a supportive setting.[41] Similarly, the structured environment of a theater has long met the neurological need to experience and resolve tension safely.

It's important to also consider that not all cultures emphasize conflict-based stories. Other structures that are non-Western resonate with their audiences relative to expected perception. *Kishōtenketsu* is a four-part narrative structure (introduction, development, twist, and conclusion) common in Japanese and Chinese storytelling. Rather than conflict, it emphasizes harmony and reflection, appealing to audiences accustomed to stories that focus on character growth, relational dynamics, and poetic resolution. In India, the Rasa Theory centers on evoking specific emotional responses, such as love, sorrow, or heroism, in the audience. This aligns with cultural traditions that prioritize art as an immersive emotional and spiritual experience rather than focusing on linear plot progression. Yoruba oral story structures use cyclical storytelling, proverbs, and audience participation to emphasize the interconnectedness of humans, spirits, and nature, resonating with West African audiences who value communal engagement and moral lessons conveyed through shared storytelling.

It's often said that good stories "show" instead of "tell." This phrase reflects a deeper neurological truth. Stories that incorporate imagery engage the brain on a multi-sensory level. Our brains are wired to respond more strongly to vivid, sensory-rich experiences than abstract descriptions. In just 13 milliseconds, the human brain can process and interpret an image—equivalent to the speed of 75 frames per second.[42] This rapid processing power allows visuals to engage us on an instinctual level, cutting through cognitive barriers that might slow down textual or verbal communication. Imagery stimulates multiple brain regions, enhancing both comprehension and retention. But our brains create immersive narrative experiences even when the input is only verbal or textual by engaging the brain's default mode network (DMN).[43] Most active during introspection and imagination, the DMN works with other regions of the brain to construct internal mental simulations, connect stories to memories, support self-reflection, and integrate story elements into a cohesive internal experience.[44]

Even after the final scene or chapter, the brain continues to process the story, often drawing connections to personal experiences or emotions. Neural connections strengthen around emotionally charged moments, making stories key drivers of long-term memory and self-reflection. Our brains select significant or emotionally impactful events to remember, while omitting less resonant details. They create causal connections between events, even when none exist, in a quest to understand how one occurrence leads to another. The phenomenon known as "apophenia" describes our tendency to perceive connections or patterns between unrelated things. This internal narrative construction turns sequences of events into narratives that reflect and reinforce our experiences and sense of self. It's why people can see the same content and have a different experience or opinion of it. Shakespeare's *Hamlet* can be seen as a tragedy of inaction by one person, while someone else might interpret it as a story about morality. Long after the story concludes, the emotional and behavioral echoes can persist, influencing how we interact with the world and the people around us, and shaping our culture and values. Shared story experiences also activate social neural networks, promoting group bonding and collective identity.

Perhaps Aristotle was not entirely wrong in thinking of plot as the "soul" of a story. As our minds actively engage with stories, we meld events together with memories, fantasies, and new sensory input. But while a plot may provide a semblance of structure, the true power of a story lies in the dynamic interplay between its narrative and the brain's desire to comprehend, feel, and make sense of our experiences. This neural engagement is what makes stories powerful tools across many types of communication, education, and entertainment. And embracing the underlying, often subconscious, processes of our neurological systems can allow creators to build systems that dynamically reinforce a pattern rather than following a prescriptive plot.

ENGAGEMENT AS MEASURABLE

By and large, determining if a story that "works" has thus far been a matter of observation. We might look at the box office numbers on opening weekend or check if it's on a bestsellers list. Or we might cite the way audiences respond using metrics like applause, replays, or social sharing. These behavioral indicators, however, are influenced by numerous variables that undermine their reliability: audiences may applaud out of politeness rather than true engagement and people often replay content as background noise while multitasking. Self-reported surveys also contain inherent biases that affect their accuracy: respondents may not correctly remember their emotional state during the event (recall bias), the phrasing of questions may influence the type of answers given (leading question bias), and people with the strongest opinions, positive or negative, are most likely to take surveys (response bias). Even so, individual reactions to a story vary widely, particularly in theater where the presence of an audience impacts how others perceive the show. In the absence of more precise measurement tools, these flawed metrics have been used for centuries.

In order to truly substantiate "engagement," it would be ideal to obtain objective, quantitative measurements while the story is being experienced.

It turns out that emotional responses manifest in the body in physical ways that are unbiasedly measurable—and the tools for observing engagement of audiences in real-time are growing increasingly sophisticated. Functional magnetic resonance imaging (fMRI) is a non-invasive technique that observes the brain's activity by detecting blood flow and oxygen levels in response to stimuli, such as key dramatic moments. Through fMRI, scientists have observed that when we are engaged in stories, certain regions of the brain become active. Electroencephalography (EEG) is a technique that records electrical activity in the brain. By measuring brainwave activity, studies using EEG have provided insights into how stories captivate and influence our minds. Cirque du Soleil's collaboration with neuroscientists in 2018 explored the effects of awe during live performances of the "O" show in Las Vegas. The results revealed significant physiological reactions, such as changes in heart rate and skin conductivity, both associated with awe. Specific brain activity patterns related to awe were recorded, including activation in the medial prefrontal cortex, which is involved in emotional processing and self-reflection, and the posterior cingulate cortex, which is linked to self-transcendence and a sense of connection to something larger than oneself.[45]

However, these methods have limited applications. fMRI requires participants to lie still in a confined and noisy machine, a setting that is far removed from a typical viewing environment, and EEG machines place small, metal discs on the scalp that are attached to a computer. The Cirque study, for example, involved 282 audience members who wore EEG caps to monitor brain activity, along with sensors tracking heart rate and skin conductance. Neither are optimal for large-scale implementation.

More recently, neuroscientist Dr. Paul J. Zak's research has advanced our understanding of how stories influence the brain through the measurement of neurochemical responses. Zak's pioneering studies developed a way to measure the release of oxytocin and identified it as the neurochemical responsible for empathy and narrative transportation. His book *Immersion: The Science of the Extraordinary and the Source of Happiness* explains how, by measuring neurochemicals in real time, creators can predict engagement and design more resonant content. Decades ago, this discovery required drawing blood from research participants in a lab. It can now be done with wearable technology. Zak has developed methods that use smartwatches and fitness trackers to capture physiological signals and provide millisecond-by-millisecond data on story engagement.

Modern neuroscientific tools are able to measure the invisible with surprising accuracy. Quantifiable neurological phenomena that resonate across all boundaries of content, genre, culture, and geography offer a definitive, measurable data of engagement that transcends subjective interpretation. This "emotional-neurological state" is reflected in the measurable brain activity and physiological responses that correspond to emotional and cognitive engagement, providing an objective indicator of how deeply an experience is resonating.

Future technologies like wearable devices that access brain activity could enable narratives to adapt in real time to audience emotions and thoughts. Brain-computer interfaces could merge cognitive responses directly with narrative progression, offering storytelling experiences that are as complex and dynamic as human thought itself.

To create an interactive experience with a story that dynamically delivers an emotionally effective dramatic arc, we need to understand two things: the ideal progression of the story—a path that leads the audience through tension, discovery, and resolution—and the guest's mental and emotional state. By knowing the state of both in real time, we can identify the delta between the guest's current engagement and how the story can proceed optimally. The system would then generate content to bridge that gap, creating moments that resonate emotionally, build on the guest's existing engagement, and guide them through a satisfying narrative experience. By adapting story elements to meet the guest's state, the system can ensure each interaction feels meaningful and coherent, sustaining attention and delivering the fulfillment the brain craves from a well-told story.

RELATIVE LINEARITY

Linearity is a powerful storytelling tool, providing a format that allows complex ideas, emotions, and events to unfold in a way that resonates with our natural experience of time. But in reality, our own path moving through life is just one slice of a vast, interconnected web of moments happening all at once. What we experience in our lives is just one perspective—a selected thread through an infinite array of events occurring simultaneously. Each person, each place, and each action we don't see exists in parallel, contributing to the world in ways we will never fully grasp. Our individual perspective on a situation may make a political figure, event, or color resonate differently than it would with someone else. Similarly, the experience of a narrative that is expansive beyond a singular progression, then, is relative and shaped by our position within it.

This realization shifts our understanding of narrative: rather than a fixed chain of events, a story becomes a vast, dynamic system where linearity is relative to the guest's chosen path through it. We're not only perceiving a sequence; we're inhabiting one version that contains infinite possibilities, where every moment has the potential to diverge, shift, or mirror countless others we may never see. Stories, then, are only linear in relation to us, while in essence they are boundless—open to endless perspectives, each contributing to the larger narrative whole.

Yes, the sequence of events is still linear in that it is *perceived by the observer as linear*—one moment after the next is the only way that we humans experience the world—but now the linearity is relative to each person and localized, rather than being universally exhibited en masse. "Narrative" becomes a personalized sequence of events. How we experience that sequence is a matter of structuring the events along a path of optimal emotional-neurological engagement.

QUANTUM COMPONENTS

If the primary function of dramatic narratives is engagement through story components, what are those components and how are they organized to achieve optimal emotional-neurological states? Aristotle identified a hierarchy of six components of plot, character, thought, diction, melody, and spectacle. How the components of drama combine to create powerful stories has been of particular interest for modern theorists and critics, too. Screenwriting lecturer Robert McKee emphasizes the importance of an inciting incident followed by a clear plot arc in his influential book *Story: Substance, Structure, Style, and the Principles of Screenwriting*. Augusto Boal's Theatre of the Oppressed theory identifies dialogue as a crucial dramatic element. In his book *The Empty Space*, director Peter Brook argues that the space itself is a crucial component, as the dynamics of the performance area influence the engagement and experience of the audience. Pauline Kael, a renowned film critic, believed that a story's emotional resonance was paramount, often assessing how well a director's vision aligned with the story and how effectively the actors conveyed emotional truths.

Clearly engagement doesn't stem from any single aspect, but rather from multiple components working together—the synergistic interaction of these components serving as the basis for creating meaningful dramatic narratives. With the understanding that story is an emotional-neurological experience, we can now reexamine their meaning and consider a system more appropriate for interactivity. The phenomenon of story, once confined to discrete forms like linear narratives and singular plotlines, might be reimagined as a dynamic, interconnected ecosystem where each contributes to the overall effectiveness of the story. This means that no single element is inherently superior; rather, each gains its importance from how it operates in relation to the others, adapting to the nature of the story and the desired audience response.

This concept mirrors principles in physics, particularly in how components interact to produce observable phenomena. In physics, nothing exists or acts in isolation. The behavior of the universe emerges from the interplay of various influences and dynamics. In physics, systems are studied from the outside in, revealing how components respond to external forces and constraints. This systems-based perspective naturally aligns with the design of interactive narratives, where characters, settings, and events evolve in response to guest input and shifting contexts.

By adopting physics as a guiding framework, interactive narratives can better adapt to the intricate, often unseen forces that shape human experience. Just as physics maps probabilities and interactions to predict outcomes, narrative design can apply similar models to chart the delivery of engaging stories.

On a very simplified level, five foundational components work across all areas of physics, explaining the universe as we know it. Matter is the substance of the universe—it includes everything that has mass and occupies space. Energy is the "ability to do work," driving all physical processes and interactions. Spacetime is the framework in which these interactions unfold, combining space and time into a single, dynamic entity. Within spacetime, fields are regions where forces act, mediating the interactions between matter and energy. Forces are influences that cause matter to move, interact, or change, arising through the behavior of fields. These components define the nature of reality, from the smallest particles to the vast cosmos—and perhaps, in their interplay, also interactive narratives.

These components are not independent like Aristotle's prescriptive components in *Poetics*, which were categorized as discrete elements, each serving a distinct function within the structure of a story. Instead, the foundational components of the universe—matter, energy, forces, fields, and spacetime—are deeply interdependent, constantly influencing and shaping one another. Matter and energy are exchangeable, as shown in Einstein's equation $E=mc^2$, which explains that mass can be converted into energy, such as in nuclear reactions, and energy can transform into matter, as seen in particle creation. Forces also mediate energy; energy is stored in fields, shaping spacetime. These components cannot be disentangled from one another.

Of course, each of these components encompasses highly intricate mechanisms and nuanced relationships. Matter, for example, ranges from subatomic particles to massive celestial bodies, each interacting with forces and fields in unique ways. Energy exists in diverse forms, from kinetic to potential, thermal to electromagnetic, and can transfer, transform, or be stored under specific conditions dictated by quantum and classical laws. Rather than diving into the granular details of particle physics or thermodynamics, we'll treat these components as broad concepts that shape and influence systems. This allows us to explore how they interact and drive larger processes without becoming lost in the specifics. By maintaining this broader perspective, we can better understand the high-level roles they might play in interactive narratives.

Forces in physics invisibly constrain and propel movement. The equivalent of an invisible but defining force in a story is the *theme*. A strong theme has a gravitational pull that draws all elements—characters, plot points, setting—into alignment, unifying them under a central idea. In stories, themes provide the overarching ideas, influencing how the story is crafted and perceived. Subtly, it guides the flow and structure. It keeps the narrative grounded, holding each piece in place so that no matter how many subplots or perspectives there are, the story feels whole. Like the pull of gravity, the theme quietly organizes, giving the story depth and purpose as every decision, interaction, and twist moves closer to the thematic core. Multiple forces can exist and interact across different levels of matter. Gravity strongly influences large-scale structures like planets and stars, while electromagnetism primarily affects particles at subatomic levels. Likewise, layers of themes can coexist in a story, enhancing or offsetting each other. This is where a story's depth is contained—our ability to derive a higher meaning from a sequence of events. Familiar themes such as the struggle between good and evil, coming of age, or love and sacrifice shape characters' journeys and drive the emotional arc.

Spacetime is our framework. It might seem obvious that the narrative approximation of "spacetime" would be things like

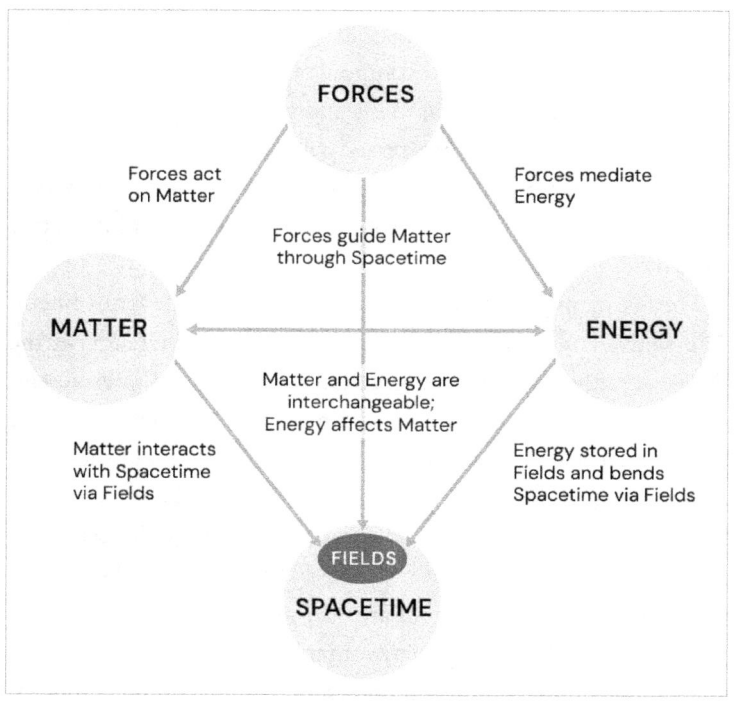

FIGURE 16. High-level conceptual diagram of the foundational relationships in physics.

the setting, the environment, or the time period. But remember that spacetime is fluid, a fabric of reality that can be changed and warped by matter, energy, and forces. It is invisible, calculable only by observing its effects. In this new domain, the arena is not the place in which the story takes place but the invisible story structure itself. If we think of spacetime as *dramatic flow*, it becomes clear that the narrative's structure isn't just the sequence of events, but is also the emotional and psychological progression that aligns with how the audience experiences time. In this sense, the audience's perception of story time—the build-up, climax, and resolution—mirrors how spacetime bends and warps under the influence of forces like theme, tone, and engagement.

Whereas traditional frameworks treat structure as a static map, conceptualizing spacetime as "dramatic flow" in the setting of a quantum narrative acknowledges that the experience of a story is fluid. This allows moments of tension, anticipation, and catharsis to be shaped by the interaction between audience energy and the story's thematic core. In this model, the audience isn't just watching a preset plot—they are actively bending and reshaping the narrative fabric.

Fields within spacetime are where energy is stored and through which matter interacts. The force fields of science fiction often seem far-fetched, but conceptually they are rooted in actual physics and already exist in the form of magnetic and electric fields that attract or repel certain materials without physical contact. Before they were scientifically understood, their mysterious abilities were considered magical or supernatural. Emotions like love—though deeply personal and mysterious—are already being unraveled by science in ways that feel just as inevitable, even if we want to resist the idea that it can be fully explained.

Fields are the *dynamics* that shape world-building. They act like the invisible hands that sculpt the environment, providing the rules, tension, and interaction that form the backdrop of a universe. But they aren't passive; they actively shape matter, forces, and even the fabric of spacetime itself. They mediate forces by translating themes into meaningful interactions, creating depth and resonance. Dynamics also organize energy by channeling audience engagement and character decisions into structured consequences. They shape spacetime by influencing how the story's structure unfolds.

For example, in a mystery story, the *genre dynamic* defines the parameters of tension through clues, red herrings, and escalating stakes, engaging the audience in the process of uncovering the truth. The *cultural dynamics* reflect societal values like justice, secrecy, or truth, shaping the story's backdrop, which might be a small town steeped in mistrust and gossip. *Social dynamics* influence interactions between characters, such as strained relationships, personal rivalries, or fragile alliances among suspects, all of which add layers of intrigue. Inner states are driven by *psychological dynamics*, like a detective's obsessive determination to solve the case or a suspect's guilt causing hesitation or avoidance. And *probability dynamics* might govern possible outcomes—shaped by character choices, unexpected twists, audience interpretations, and even randomness, creating an engaging and unpredictable plot. Together, dynamics layer and interact to shape the story and guide the audience through an engaging experience, much like how different types of fields in physics interact.

Matter in physics is a scientific construct, strictly defined by its physical properties and behavior in the universe. In quantum narratives, matter should not be limited to the purely physical. Instead, it should encompass everything tangible or perceptible within the story world. Matter, in our story parallel, includes all *sensory forms* such as characters, dialogue, movement, and audio-visual elements, even though these aren't "physical" in the traditional sense. These sensory forms shape how the guest perceives and navigates the world, and can be intentionally designed for particular effects. For instance, dialogue choices can shape the plot by influencing alliances, creating ripples that feel as "real" as moving an object. Atmospheric lighting or sound distortions can act as barriers or guide where guests move. By expanding the concept of matter in this context, creators can use every sensory input from music and lighting to sound effects and more to actively shape the story. In this way, they become tangible aspects of the story world.

What, then, is **Energy**? Is it the conflict? Some stories engage without direct conflict, so it must be something broader. Recall that energy drives the transformation and movement of matter. In this model, the audience or *guests* act as energy; they are the driving force that powers and activates the story. Just as energy initiates motion or transformation in physics, the guests' engagement through choices, focus, and reactions meaningfully propels the story forward. Guest participation provides the momentum that keeps the narrative dynamic and responsive. Without their interaction, the story remains static, like potential energy waiting to be released. The guests interact with the system, influencing matter (characters, environments), interacting with fields (dynamics), and reshaping spacetime (dramatic flow). Their participation is what uniquely animates the narrative system.

This structure for quantum narratives is a compelling approach because it aligns with the principles of emergent systems and contemporary understandings of engagement and cognition. Unlike

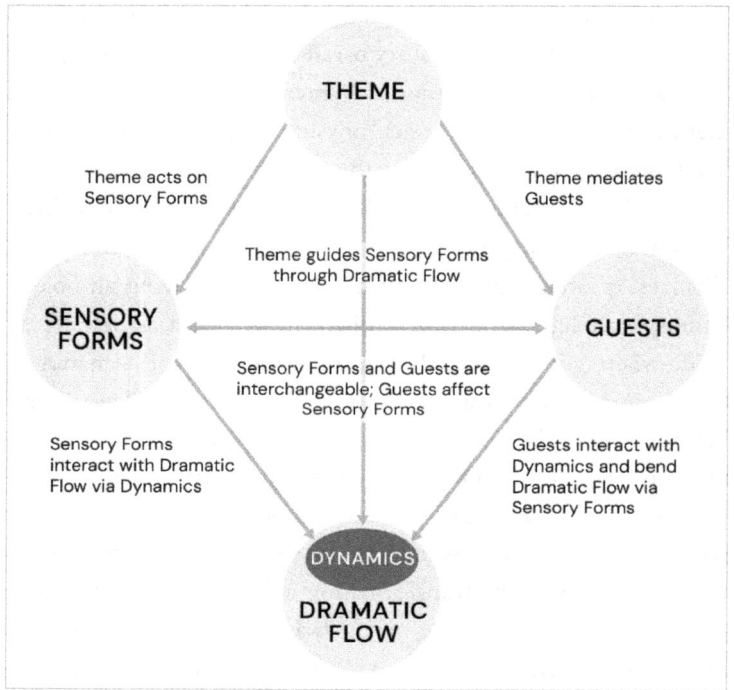

FIGURE 17. High-level conceptual diagram of the foundational relationships in quantum narratives.

the framework offered by Aristotle's *Poetics*, which does not account for audience interaction and requires abstraction from the natural world to control the story, quantum narratives harness an organic model to foster distributed agency, allowing story elements and guest participation to form a simultaneously cohesive and adaptive whole. This actually mirrors the way modern media consumption prioritizes interactivity and personalization, as seen in gaming mechanics, virtual reality experiences, and even social media algorithms. But since quantum narratives provide a fluid structure responsive to our neurological processes, deeper emotional and cognitive engagement will ultimately be possible.

Stories released from classical frames become ecosystems of meaning through interaction. The story itself ceases to be a mere *telling* and occupies a new domain of the heightened narrative experience of presence.

WHERE DOES THAT LEAVE *US*?

Discussions around people who engage in interactive, dynamic and spatial narratives describe them as "players," "audiences," "participants," or "users." The respective terms reflect the expected roles of a person experiencing the narrative across gaming, entertainment, location-based environments, and software, respectively. However, none of these terms fully captures the fluid, adaptive nature of our involvement. Interactive designer and director Jeff Wirth dubbed this new category of audience "spect-actors" in his book *Interactive Acting*. Rather than passive spectators, he views them as active contributors to the story.

In quantum narratives, I prefer to think of those who join the experience as guests. This word is uniquely fitting because it reflects both the agency and adaptability inherent in these dynamic storytelling forms. Guests are welcomed into an experience rather than confined by rigid expectations. Like entering a host's home, guests navigate a space where their actions influence the environment, yet they are not entirely in control. A guest can choose their level of engagement, choosing to "skim, swim, or dive" deeply into the experience. Immersive designers commonly refer to people who passively observe an experience as skimmers, whereas swimmers explore its details, and divers submerge themselves fully in its depth and complexity. And, like in life, there may be moments during the experience when guests are quietly listening and others when they are actively engaging. Guests in this context are not merely passive observers or mechanical walk-through participants; they are people whose presence in the experience matters to the evolution of the story.

The term "guest" also acknowledges the hospitality and intentionality of the narrative space. It suggests a shared experience where creators design with care, crafting environments and systems that adapt to diverse levels of engagement. Guests are invited to explore and influence without the pressure of predefined roles, making their journey personal and meaningful. In this way, quantum narratives respond to both design and choice.

This alters the nature of authorship significantly. Classical authors internally process the rules and structures of storytelling and deliver fixed content for the page, stage, or screen. With quantum narratives, authors will need to confront the difficult challenge of externalizing those invisible understandings into algorithms and conditions that then produce unpredictable content. The focus and purpose of an author in this new domain shifts from delivering a definitive work to shaping the boundaries, possibilities, and tendencies of generative systems. In this sense, authorship is akin to the role of a system designer in that the end result of their contributions is less about a concrete end product and more about shaping the parameters of an experience. The final output might even surprise them.

As with any evolving medium, this shift will inevitably affect us as a society. Initially we expect familiar formats, like our skimmers who passively observe interactive content as if it were in a classical medium. However, people historically adapt to changes, a crucial component in in the evolution of story. For example, the shift in film editing techniques, particularly the adoption of rapid editing, has significantly influenced audience engagement and comprehension. Cognitive research indicates that viewers have adapted to editing styles over time, enhancing their ability to process faster-paced narratives. Studies have shown that changes in the visual field, such as quick cuts and increased motion, naturally attract viewers' attention and eye movements, facilitating their adaptation to modern editing styles.[46] Additionally, the concept of "intensified continuity" in contemporary Hollywood films, characterized by more rapid editing and closer framing, reflects a shift in filmmaking that aligns with audience expectations and cognitive processing capabilities.[47] These findings suggest that as filmmakers have embraced novel editing techniques, audiences have correspondingly developed the cognitive skills necessary to follow and comprehend these faster-paced narratives effectively.

Audiences accustomed to the polish of high-budget films might initially find the rawness of user-generated content on platforms like YouTube less appealing, but over time, they can come to appreciate authenticity and personal connection as much as technical quality. Platforms like Twitch, where streams like "Twitch Plays Pokémon" involve real-time interaction, generate emergent stories through audience participation. The story unfolds dynamically based on the collective actions and decisions of the viewers. It *feels* like a story even if it lacks a traditional plot structure.

In times of change, humans do what we have always done: evolve. This adaptability not only allows individuals to navigate novel formats but also pushes creators to innovate, fostering a cycle of mutual evolution between the guest and the medium.

THE NEW DOMAIN

Wrapping and revolving around each other with inevitability like the twin strands of DNA, the double helix of science and story run in tandem, each lending depth and perspective to the other. While science deciphers the code of the universe, revealing its mysteries, story breathes life into those discoveries, expanding the human experience with emotion and imagination. And just as DNA constantly evolves, carrying the vibrant story of life, the synergy between science and story evolves too, sparking endless wonder and curiosity. Together, they spiral forward, a testament to and a tool for humanity's unending quest for understanding and connection.

Technological advancements give us new twists—like the telescope revealed secrets of the universe amongst the stars, leading to the forsaking of creation stories, and cathode ray tubes unveiled a subatomic domain so abstract and strange that it compelled the embracing of story for the purpose of comprehending the world once again. Today's emerging digital tools are likewise having a profound impact on both science and story—artificial intelligence, game engines, and virtual reality platforms are creating unprecedented possibilities. As we continue to spiral forward, it becomes increasingly clear that the boundaries between science and story are not fixed, but fluid, and the next phase of narrative evolution may involve a fundamental rethinking of how we perceive both, together.

Clinging rigidly to classical storytelling models in interactive narratives is like trying to map quantum particles with a ruler. What worked for the visible domain falls apart when the rules of engagement shift. The foundations of story must evolve just as science has. Radically, if needed. Moving beyond traditional deterministic storytelling methods to a probabilistic, systems-based approach rooted in physics may seem unconventional, even strange, but it is increasingly becoming a necessity. Grounding interactive narratives in principles from physics, neuroscience, and complex systems doesn't just provide a system; doing so bases stories dynamism in patterns that resonate with our cognitive and emotional-neurological tendencies.

In this new domain, we shift our focus onto the invisible probabilistic conditions rather than just the visible deterministic expressions. We advance the story components from being the product of manual construction to being dynamically reactive systems. And we reprioritize interactivity and uncertainty, previously implemented as a feature, to being fundamental.

The basic framework of quantum narratives explored thus far has focused on the structural core of the story experience and system, but it is the experiential layers that give it life, texture, and emotional weight. These are equally vital for creating an engaging and resonant interactive narrative and worthy of another volume of exploration. After all, narrative is not just how the audience interacts with the story, but how deeply they feel it. Immersion pulls the audience into the world, dissolving the boundary between the narrative and their reality. This can manifest through rich sensory design, environmental storytelling, and responsive systems that make the world feel alive and reactive. Relatable characters, meaningful stakes, and personal investment are the factors that compel audiences to care about the unfolding story and feel the gravity of their decisions.

The classical domain has so successfully deployed experiential elements that they remain relevant, although perhaps in an evolved manner. Designers might develop systems that change dynamically to diverse audience needs, where content delivery (such as language preference, visual representation, or interaction speed) adjusts based on audience needs. Accessibility ensures that the narrative is not limited to a select few but can resonate broadly across different abilities and experiences. By designing for inclusivity—whether through adaptable interfaces, multiple modes of interaction, or narrative pathways that account for different playstyles—creators ensure a more diverse audience can connect meaningfully with the story. This inclusivity isn't just about functionality; it deepens the narrative by allowing more perspectives to shape and experience it.

These sensory layers help creators to bridge the gap between technical systems and human perception. A well-crafted structure might guide the guest's journey, but it is the feeling of presence, the emotional highs and lows, and the seamless accessibility that truly define the experience. When these elements work in harmony, the narrative becomes more than just a series of choices. It becomes a living, breathing world.

To further advance specific elements of narrative systems like believable characters and emotionally compelling plots, it will be essential to involve disciplines beyond narrative and computer science, who have been the primary explorers thus far. Behavioral psychologists can contribute insights into how humans process emotions and make decisions. Dialogue that reflects cultural and social nuances might come from sociolinguists, while ethologists bring expertise in instinct-driven behaviors for designing believable agents. Game theorists can guide the design of interactive narratives, ensuring choices feel meaningful and coherent, and complex systems theorists can provide tools for crafting dynamic, emergent storylines where interconnected elements produce unpredictable yet cohesive outcomes. In addition, artificial intelligence ethicists can guide decisions on narrative agency and machine-driven adaptability, while neuroscientists can offer a deeper understanding of cognitive engagement and emotional triggers. Design thinkers can ensure that user experiences align with interactive storytelling goals, and media studies experts can bring insights into how audiences consume and interact with stories across different platforms. Collaborating across disciplines, we can push the boundaries of narrative systems, creating stories that resonate deeply both emotionally and intellectually. Consider this to be an open invitation to all who wish to take part in shaping the future of narrative.

THOUGHT EXPERIMENTS

To make the concepts of quantum narratives more concrete to the largest audience possible, we can draw from the approach physicists use to grasp abstract ideas: thought experiments. The following examples serve as initial explorations into how guest presence and agency might dynamically influence narrative form. Through these exercises, we begin to unravel the principles that underpin adaptive storytelling, where the boundaries between audience and narrative dissolve, and the act of observation itself becomes a catalyst for change.

THOUGHT EXPERIMENT #1:
The Adaptive Prop

Imagine a single prop, like a chair, in the center of a scene. Its purpose and meaning evolve based on guest interaction. If a guest chooses to focus on the chair and examine it, it becomes a critical artifact in the narrative—a throne that signifies authority or a relic tied to the story's lore. If the guest ignores the chair and shifts their attention elsewhere, it fades into the background, serving only as part of the set. Should a guest interact directly—perhaps by sitting on it—it transforms into a key narrative device, sparking a scene where characters react to this newfound role of the guest. The chair's function and significance are not predetermined but shaped by the choices and focus of the guest.

In this new domain, guest presence is meaningful and consequential. Plot, characters, and themes adapt to their presence, creating a story that is both structured and ever-changing, resonating with the complexity and interconnectedness of life itself. By interacting with the story, the guest affects how events unfold. Their choices, timing, and focus act like observations where audience agency and story structure exist in balance.

THOUGHT EXPERIMENT #2: The Dual-Character Mask

Imagine an actor wearing a mask that changes expressions depending on who is looking at it. To one guest, it might appear joyful, while to another, it could seem sorrowful. This mask embodies multiple potential emotions simultaneously, with its final expression determined by the observer's perspective.

Responses to interactions within the story are not predetermined, but instead are dynamic, adapting in real time to the experience of the guest. These changes are relative to the specific emotional-neurological sequence activated as the guest engages with it.

THOUGHT EXPERIMENT #3:
The Quantum Sand Timer

Imagine a sand timer on stage that doesn't measure time in a traditional sense. Instead of sand falling uniformly, each grain represents a different potential moment in the story. Sometimes the sand falls quickly, skipping over parts of the narrative, and other times it moves slowly, stretching moments out. Occasionally, the sand flows upward, reversing time and allowing the characters to revisit past events with new perspectives.

Plot is not a preset sequence in this domain. Instead, plotlines emerge from probabilities and perception. Different point of views, timelines, and arcs can exist simultaneously and in various states, holding space for alternative paths, perspectives, or even realities.

THOUGHT EXPERIMENT #4:
The Rippled Mirror

Imagine a large mirror on stage, but instead of reflecting a direct image, it ripples and distorts based on the actions and choices of the characters. If a character stands still, the mirror shows an unbroken, clear reflection. If they take an action, the mirror ripples outward, revealing shifting

> *images of other characters, fragmented moments from parallel plotlines, and subtle hints of events yet to come.*

Story components are persistently interconnected across timelines, plotlines, and even between guests, forming a network where one guest's choices or interactions can reverberate through shared elements such as environments and characters.

OUR QUANTUM FUTURE

Reconceptualizing the foundations of our stories offers exciting possibilities for the future of narrative. Imagine stories functioning outside of our control of time and space, where emotional resonance is a result not of not being told something, but rather of being experienced in the context of our own personal journeys. It is more encompassing than what has been proposed in Janet Murray's *Hamlet on the Holodeck* and Henry Jenkins' *Convergence Culture* or imagined in talks from classical visionaries like Neil Gaiman and Steven Spielberg. In a futuristic vision, adaptive narratives might respond to guests' thoughts and emotions in real time, seamlessly connecting their consciousness to a narrative framework. As our future guests enter a quantum narrative, the world around them shifts and adapts, forming a story uniquely tailored to their innermost desires, fears, and choices.

Boundaries between author and audience dissolve. The narrative is symbiotic between system and human, a continuous dance of interaction and adaptation. As guests think, feel, and react, every decision and emotional response feeds into the narrative algorithms, crafting a story that is deeply personal and ever-evolving. The result is an immersive experience that feels as real as life itself, yet symbolically heightened and dramatically crafted.

In one moment, a guest might find themselves in a bustling futuristic city, navigating political intrigue and uncovering hidden truths. In the next, they might be transported to a serene, ancient forest, where they witness mythical combats that confront their own inner demons. The narrative flows seamlessly, guided by both the guest's subconscious and the system designer's overarching vision. It's a living story that breathes with the participant, offering a unique journey every time. It can unfold in solitude or be shaped collectively, where gatherings of guests convene to evolve the narrative together with each part being perceived differently based on where they are in the timeline, environment, or social construct.

However, amid ever-advancing technological marvels, it is crucial to remember the heart of storytelling: us. No matter the technology or the framework, humans remain at the center of the narrative experience. It is the human who interprets the story, and brings magic and meaning. While technology can facilitate a dynamic interaction, it is the human element—the guests' emotional responses, the collective consciousness of the group, and the shared human experience—that brings a story to life. This dynamic network of technology, humanity, and story is what makes the future of narrative exciting and full of potential. The journey into the quantum narrative paradigm is not just about new technologies or storytelling methods; it is about rediscovering the essence of what makes stories meaningful.

Classical storytelling frameworks have traditionally centered around a single individual's perspective, often designed to benefit and highlight one protagonist. This approach mirrors a linear, cause-and-effect view of life, where the world seemingly orbits around the experiences of one character. It reinforces the notion that our lives, like stories, must be driven by personal desire, conflict, and triumph.

This singular focus reflects a self-absorbed view of the human experience, suggesting that meaning and transformation stem primarily from the journey of the individual rather than the collective. It echoes cultural values that prioritize personal agency, heroism, and individual achievement over shared experiences, communal struggles, or environmental consequences. Over time, this emphasis on individualism has seeped deeper into societal structures, influencing everything from education and career paths to relationships and social movements.

Modern cultures, particularly in the West, have increasingly framed success and fulfillment as personal conquests—defining worth by self-realization and the ability to stand out from the crowd. The obsession with individual narratives fuels industries built around self-help and influencer culture, reinforcing the belief that each person is the protagonist of their own saga, with others relegated to supporting roles.

As this worldview tightens its grip, the idea of collective identity often fades, reducing empathy and minimizing the importance of shared responsibility or mutual growth. Societies driven by this model may unintentionally foster disconnection, encouraging competition over collaboration and reinforcing the myth that solitary achievement holds the greatest significance. Just as storytelling has become a reflection of this cultural shift, the stories we consume, create, and value continue to reinforce the cycle, shaping how we perceive both ourselves and the world around us.

By embracing a new narrative paradigm, we are not merely expanding the possibilities of story; we are re-exploring our humanity from a more connected perspective. While the story experience remains a personal journey—one we navigate linearly through our limited point of view—the narrative itself becomes part of a larger, interconnected web that extends beyond the individual.

This shift acknowledges that even as we move through stories from our unique vantage point, the paths we take are intertwined with those of others. Our choices and actions unfold within a broader context, reflecting the reality that personal transformation rarely happens in isolation. As the Marvel Cinematic Universe has begun to explore, every protagonist exists alongside other stories in progress, and these overlapping threads influence and intersect, sometimes even redefining the journey we believe to be ours alone.

Narratives can still feel personal while subtly revealing the larger forces at play if the guest arc is treated as vitally as a character's development. If an unfolding story offers glimpses of a more intricate, communal reality, the narrative will not just reinforce personal growth but also foster a deeper awareness of how our lives, no matter how singular they may feel, are part of something profoundly larger and richer.

Just as quantum mechanics revealed that our perception of time and space as separate entities is an illusion, this new paradigm fosters a deeper understanding of our shared human experience, urging us to see the beauty and complexity of our shared journeys. It challenges us to move beyond solitary heroism and invites us to acknowledge that true heroism lies not in personal conquest but in the strength of our connections, the empathy we extend, and the mutual influence we share.

In embracing this transformative model, we are called to transcend the confines of individual-centric plots and idealized characters and instead embrace our shared humanity within the larger universe. Our stories are not merely personal expressions but part of a greater whole. Quantum narratives offer us a dynamic framework to explore our collective journey and the invisible connections that define our existence through the power of story.

EPILOGUE

A PLAY ON POSTULATES

Recall how physicists use theoretical constructs such as the blackbody to observe and measure abstract concepts and create models that predict how objects emit and absorb thermal radiation at different temperatures. This abstraction allows scientists to understand real-world phenomena, such as the behavior of materials at high temperatures, by comparing them to the blackbody ideal. Similarly, a black box theater is a minimalist and versatile space used by artists to experiment with performances in a controlled environment. The simplicity of the black box eliminates distractions, enabling creators to focus on the essence of the performance and innovate with staging, sound, and movement. Both the blackbody box and the black box theater enable the exploration and development of new ideas and applications in their respective fields.

A game engine is like a virtual black box theater—a neutral platform for creative expression and experimentation in the digital realm. Just as a black box theater provides a simple, customizable space, a game engine offers developers the basic foundation to build within, including tools for essential functions like rendering graphics, simulating physics, and managing audio. This virtual environment supports explorations that can range from crafting learning environments worlds to developing collaborative gameplay mechanics.

A software engine with a framework designed to support narrative interactions rather than physical ones might prioritize tools and features that support rich storytelling. Instead of simulating realistic physical interactions, the engine might provide robust support for managing our new narrative elements, ensuring that every choice or action taken by the guest can lead to meaningful consequences within the story.

Imagine this narrative engine contains the algorithms defining the forces that govern the rules and regulate the probabilistic consequences of audiences who interact with it in order to achieve the optimal emotional-neurological state within a dramatic structure. It is a virtual black box theater without the spatio-temporal or physical limitations of the natural world. Like theater walls, the engine's computational power imposes boundaries that shape the kinds of conflicts, characters, and concepts it can conjure.

Rigorous scientific scrutiny led to the formation of the quantum postulates. These postulates are the fundamental principles that define behavior at the quantum level. Let's use simplified versions of the postulates to influence the progression of events in a story that reflects the inherent probabilistic nature of a quantum narrative—a play on postulates, if you will.

EPILOGUE

I generated the following text in the story of Aria using ChatGPT4 in less than five seconds. I fed ChatGPT4 a single prompt, but for the sake of clearer discussion here about each postulate, I am presenting the prompt in parts. The relevant response from ChatGPT4 to each part of the prompt follows in italics. In full disclosure, I made two changes to the content generated to remove references to the narrative engine, which could have been avoided with a more detailed prompt. However, I did leave in the ChatGPT4's assumption of the word "audience" rather than the preferred word "guest."

POSTULATE 1:
State of the System

The state of the system at any moment captures all possible information about the system and can be represented as a vector pointing in a specific direction within an infinite-dimensional space.

Quantum mechanics postulate 1 states that the "State of the System" at a single moment in time can be thought of as a snapshot that captures all possible information about the system at that instant. The "system" refers to the physical entity or collection of entities being studied, such as an electron, a photon, a molecule, or any quantum object. It is the subject whose properties and behavior are being analyzed and measured. The "state of the system," on the other hand, represents all the information about the system at a particular moment in time. We can visualize this state as an "arrow" or vector pointing in a specific direction within an infinite-dimensional space.

PROMPT:

Based on the following description, craft a simple, consistent theatrical narrative that could play out progressively and dramatically in a few sentences at each step while illustrating the point made in the step:

In our theoretical narrative engine, the engine itself is the System. The State of the System is the audience's position within it. Represented by an arrow, everything they can possibly know about the dramatic narrative is available to them at the precise moment and position (the state) that they are in: their location, the setting, the characters. And they only know what is there, not what is in another moment or position within the infinite realm of possibilities of the virtual black box.

> As an audience member enters the virtual black box, they find themselves in a dark forest. At this precise moment, they hear the chirping of crickets, see a well-lit path ahead, and observe a character named Aria standing next to the path, looking distressed.

POSTULATE 2:
Measurement Values

Every measurable quantity is tied to a mathematical operator, and the results can only be continuous or discrete.

When you measure certain features in quantum mechanics, sometimes the results can vary freely, like turning a volume knob (this is called continuous). Other times, the results are more like flipping a switch with specific settings (this is called discrete). Discrete packets of energy, called "quanta," are unique to quantum theory and are where it gets its name.

> **PROMPT (CONT'D):**
>
> Just as classical measurements tend to be continuous, classical dramatic narratives tend to be continuous along a predictable trajectory. However, in order for our engine to be interactive, it will have to deliver data discretely rather than continuously as the audience progresses through the narrative. This will require us to break down the "tangible" (which I use to mean anything that has virtual mass, as opposed to force) into the smallest possible packets that can be meaningfully distributed over the course of the dramatic work. The narrative will need to be quantized.
>
>> *As the audience member approaches Aria, she reveals small tidbits: "I've lost something precious"—a quantum of information. As they continue along the path, other bits of narrative reveal themselves discreetly, like signposts containing parts of a poem.*

POSTULATE 3:
Observables

Observations make changes to the system being studied.

Observables in Quantum Mechanics are things about the State that we can measure. Think of each measurable feature (like energy or position) as a tool. When we use one of these tools on our system (the arrow), it can twist, turn, or reshape that arrow in specific ways.

PROMPT (CONT'D):

In our narrative engine, the audience can move about the virtual black box and interact with the virtual assets made available to them by the engine in the position they occupy at the moment of the dramatic structure they are in. Where the audience navigates within the system (our arrow) and which characters, objects, storylines, and features they engage with is registered by the system. These registrations will accumulate to affect the probability of future quanta delivery to achieve the desired neurological effect. The observables in the narrative engine are our audience and the narrative quanta.

> *Choosing to help Aria, the audience member follows a set of footprints. They peer into a shimmering pond, touch a glowing stone, and listen to an owl's cryptic message. All these interactions—the pond, stone, owl, and Aria's statements—are logged by the engine, setting the stage for future narrative possibilities.*

POSTULATE 4:
Probability and Measurement

Before a measurement, the system has a range of possible outcomes, each with a certain probability, and measurement selects an outcome based on these probabilities.

Quantum mechanics fourthly postulates that before you measure, you have various possible outcomes, each with its own likelihood. When you do measure, you roll the dice and nature picks an outcome based on those chances. The key point is that you can't predict the exact outcome beforehand, only its probability. The fundamental unpredictability and probabilistic nature of the quantum world is the concept famously likened by Einstein to a game of dice.

> **PROMPT (CONT'D):**
> When we set up our narrative engine, we provide it with elements, properties and assets as a basis for dynamically generating content. Before the experience starts, anything is possible within the realm of dramatic constraints, like the sides of the dice. As the audience engages with the narrative, the engine picks outcomes and responses based on the probability of it happening within the narrative. The algorithms for dynamically generated content don't produce the outcome beforehand.
>
>> *The audience might choose to touch the glowing stone, but its outcome isn't predetermined. If touched, the engine may probabilistically decide based on the audience's previous interactions that it's a portal stone or perhaps a clue. Aria could hint its importance, or it could be a decoy.*

POSTULATE 5:
Wave Function Collapse

A system exists in a superposition of multiple states simultaneously, and measurement collapses this superposition into a single, specific state, defining the system's exact outcome.

Imagine a spinning roulette wheel where each slot represents a possible quantum outcome, and the speed of spin indicates uncertainty. The act of measuring is like the wheel suddenly stopping, with the ball settling into a specific slot. This unexpected settling represents the wave function collapse, giving a definite outcome from the quantum uncertainties. Until you measure, the exact outcome doesn't "exist" in quantum mechanics the way we're used to thinking.

> **PROMPT (CONT'D):**
>
> When our story engine is set up, it theoretically contains all potential storylines, character arcs, and plot developments, existing simultaneously as possibilities. These are not fixed until the audience interacts, making choices or influencing the narrative direction. It's only when the audience engages—much like making a measurement in quantum mechanics—that the narrative "collapses" into a specific storyline, solidifying one path out of many potential ones. This mirrors quantum mechanics' fifth postulate, where the act of engagement determines the exact outcome, crafting a unique narrative experience from a realm of infinite possibilities.
>
> What the audience is going to do at any given moment in the narrative is uncertain until they actively interact with the system. The spinning roulette wheel of possibilities in our narrative engine "stops" or selects and activates the outcome dependent on their action. The outcome can't be known for certain until the audience interacts.
>
>> *Upon deciding to throw the stone into the pond, the myriad of narrative possibilities collapses to one: the stone creates ripples that form an image, revealing a map to a hidden treasure, sealing the narrative's direction for that particular choice.*

POSTULATE 6:
Evolution Over Time

A quantum system evolves predictably over time, continuously changing its superposition of states until a measurement collapses it into a single, definite outcome.

If left to its own devices, the quantum system (our arrow) changes over time, dancing to the tune of a specific rule called the Schrödinger equation. Think of this equation as the equivalent of an author who dictates how the system evolves—where the story goes, how fast it moves, and how it transitions between all possible states. It's like a dramatic structure that tells the quantum system exactly how to behave over time without any outside interference, describing a smooth, continuous flow of change. The Schrödinger equation doesn't predict a single outcome but rather how all potential outcomes mix and shift until something external, like a measurement, forces a decision and collapses the system into one definite state.

> **PROMPT (CONT'D):**
> The objects that have virtual mass in our engine (or are "tangible") are affected by invisible forces that act upon them. These intangible forces dictate the course of dramatic action over time, regardless of audience interaction. If the audience chooses to do nothing, these forces will still compel the tangibles along a dramatic structure towards an optimal neurological-emotional state.
>
> *Regardless of the audience's choice, the virtual world evolves. The sun sets, shadows grow longer, and the narrative darkens. Aria, propelled by her quest,*

> ventures deeper into the forest with or without the audience, leading her towards her destiny.

POSTULATE 7:
Quantum Spin

> Quantum particles have a property called "spin," acting unpredictably and existing in a mix of possible states, only revealing a definite state when measured.

Quantum particles have a quirky feature often compared to "spin." But it's not an actual, physical spin. It's more like a tiny magnet that behaves unexpectedly. In the everyday world, a magnet has a clear north and south pole. This quirky "quantum magnet" doesn't just align north or south when you try to measure its orientation; instead, sometimes it acts like it's pointing halfway between, or even in some impossible direction. This special quantum property sometimes behaves as if the particle is spinning and sometimes completely defies expectations.

> **PROMPT (CONT'D):**
> Perhaps the most unpredictable component in this entire theoretical situation is the audience themselves. The complexities of human nature drive unexpected choices, behaving erratically at times in ways that are impossible to understand, defying expectations. Yet psychologists have tried, and in doing so have developed a construct that attempts to encapsulate the "spin" through the Theory of Mind. When constructing a story, a classical author anticipates the beliefs, emotions, and reactions of their audiences, essentially creating a mental representation of the audience's mindset. Characters possess their own

motivations, desires, and secrets, allowing readers to relate to, empathize with, or even detest certain characters. Through the nuances of plot and dialogue, the system guides readers' perceptions, often leading them to infer underlying emotions or motivations that aren't explicitly stated.

As the journey continues, Aria's behavior becomes more unpredictable, echoing the erratic nature of quantum spins. One moment she's ecstatic, and the next she's overwhelmed by despair. The audience, now deeply involved, makes an irrational choice: they begin to sing an ancient lullaby that wasn't part of the known narrative quanta. This spontaneous act resonates with Aria, stirring a memory.

Aria stops abruptly, looks at the audience with tear-filled eyes, and says, "That's it! The lullaby my mother sang to me!" With newfound clarity, she connects the song's lyrics to the map's symbols. Together, with the audience leading the way, humming the lullaby, they arrive at a magnificent, hidden grove. At its center is a chest.

Aria approaches and gently opens it. Inside isn't gold or jewels, but a series of old letters and a tiny music box. She winds it up, and it plays the same lullaby. "The treasure isn't riches," she murmurs, "but memories. Memories of love, family, and the songs of our past."

The input in our Play on Postulates example was a static list, creating the structure for the narrative engine to incorporate guest interactivity into the story in real-time. A narrative engine designed to create content based on the quantum mechanics postulates would generate storylines where probabilities and uncertainties play a central role. In this engine, characters and events would not follow a linear, deterministic path; instead, their actions and outcomes would be influenced by probabilities, much like how particles in quantum mechanics do not have definite positions until measured. This could lead to a dynamic storytelling environment where multiple potential storylines could coexist until specific player choices cause certain narratives to collapse into a single, experienced reality—all without writing a single line of content, dialogue, or description *before* the experience.

NOTES

SETTING THE STAGE

1. Copernicus, N. (1543). *De revolutionibus orbium coelestium, libri VI* (p. 9v). Nuremberg: Johannes Petreius.
2. Roman Catholic Church. (1633, June 22). *Sentence of condemnation of Galileo.* Pronounced by the Holy Office of the Inquisition. Translated and archived by the University of Missouri–Kansas City, School of Law. Retrieved from https://law2.umkc.edu/faculty/PROJECTS/FTRIALS/galileo/condemnation.html
3. Whiteside, D. T. (1970). The Mathematical Principles Underlying Newton's *Principia Mathematica. Journal for the History of Astronomy, 1*(2), 116-138. https://doi.org/10.1177/002182867000100203
4. Figure 1. Newton, I. (1687) *Philosophiae Naturalis Principia Mathematica.* Londini: Jussu Societatis Regiæ ac Typis Josephi Streater. Prostat apud plures Bibliopolas. Anno. Retrieved from https://www.loc.gov/item/28020872/
5. Figure 2. Einstein, A. (1997). *The Collected Papers of Albert Einstein* (Vol. 6: The Berlin Years: Writings, 1914–1917, English translation supplement, p. 98) (A. J. Kox, M. J. Klein, & R. Schulmann, Eds. & Trans.). Princeton University Press.
6. Hawking, S. (1988). *A Brief History of Time.* Bantam Books.

A NARRATIVE PARADOX EMERGES

7. Lindbergh, B. (2013, September 26). Looking back at *Myst*: 20th anniversary. *Grantland.* Archived at https://web.archive.org/web/20130926162310/http://www.grantland.com/story/_/id/9713372/looking-back-game-myst-20th-anniversary
8. Murray, J. H. (1997). *Hamlet on the Holodeck: The Future of Narrative in Cyberspace.* Free Press.
9. Aarseth, E. J. (1997). *Cybertext: Perspectives on Ergodic Literature.* Johns Hopkins University Press.
10. Aylett, R. (2000). Emergent narrative, social immersion, and "storification." In *Proceedings of the Narrative and Learning Environments Conference (NILE00).* Edinburgh, Scotland.
11. Mateas, M. (2000). A Neo-Aristotelian theory of interactive drama. In *Proceedings of the AAAI 2000 Spring Symposium on Artificial Intelligence and Interactive Entertainment.* AAAI Press.
12. Fernández-Vara, C. (2011). Game spaces speak volumes: Indexical storytelling. In *Proceedings of the 2011 DiGRA International Conference: Think Design Play.* Author's final manuscript.
13. Millard, D. E., Oyarzun, D., & others. (2012). The narrative braid: A model for tackling the narrative paradox in adaptive documentaries. In *Interactive storytelling: ICIDS 2012* (pp. 62–73). Springer. https://doi.org/10.1007/978-3-642-34851-8_6
14. Roth, C. (2018). Ludonarrative hermeneutics: A way out and the narrative paradox. *Transactions of the Digital Games Research Association, 3*(3), 113–137.
15. VanDerWerff, E. T. (2018). Netflix's *Black Mirror: Bandersnatch* shows the limits of interactive TV. *Vox.*
16. Robertson, A. (2019). *The Under Presents* turns virtual reality into immersive theater. *The Verge.*
17. Ryan, M.-L. (2001). *Narrative as virtual reality: Immersion and interactivity in literature and electronic media.* Johns Hopkins University Press.
18. Figure 5. Wikimedia Commons. (n.d.). Euclid's *Vat ms no 190 I prop 47* [Digital image]. Retrieved from https://commons.wikimedia.org/wiki/File:Euclid_Vat_ms_no_190_I_prop_47.jpg

19. Figure 6. Mateas, M. (1997). *An Oz-centric review of interactive drama and believable agents* (Technical Report CMU-CS-97-156). School of Computer Science, Carnegie Mellon University. Retrieved from https://www.cs.cmu.edu/~michaelm/publications/CMU-CS-97-156.pdf
20. Figure 7. Mateas, M., & Stern, A. (2002). *Architecture, authorial idioms and early observations of the interactive drama Façade* (Technical Report CMU-CS-02-198). School of Computer Science, Carnegie Mellon University. Retrieved from http://reports-archive.adm.cs.cmu.edu/anon/2002/CMU-CS-02-198.pdf
21. Li, B., & Riedl, M. O. (2015). Scheherazade: Crowd-powered interactive narrative generation. In *Proceedings of the Twenty-Ninth AAAI Conference on Artificial Intelligence* (pp. 4319–4320). Association for the Advancement of Artificial Intelligence. Retrieved from https://faculty.cc.gatech.edu/~riedl/pubs/aaai15.pdf
22. Figure 8. Riedl, M. O., & Young, R. M. (2010). Narrative planning: Balancing plot and character. *Journal of Artificial Intelligence Research*, 39, Figure 15, p. 259. Retrieved from https://faculty.cc.gatech.edu/~riedl/pubs/jair.pdf
23. Riedl, M. O., & Young, R. M. (2010). Narrative planning: Balancing plot and character. *Journal of Artificial Intelligence Research*, 39, 217–267.

CLASSICAL CATASTROPHES

24. Bratton, G. (2017, January 13). Dissecting a choose-your-own-adventure book. Gregory Bratton: *WordPress Blog*. Retrieved from https://gregorybratton.wordpress.com/2017/01/13/dissecting-a-choose-your-own-adventure-book
25. Rubin, P. (2018, December 28). How the surprise new interactive Black Mirror came together. *Wired*. Retrieved from https://www.wired.com/story/black-mirror-bandersnatch-interactive-episode/
26. A seed in *Minecraft* is like a code that tells the game how to create the world. When you start a new game, *Minecraft* uses this code to decide what the world will look like, such as where mountains, forests, and

caves go. If you use the same seed again or share it with a friend, you'll get the exact same world. If you don't pick a seed, the game just makes up a random one, so every world looks different.
27. RimWorld Wiki contributors. (n.d.). AI Storytellers. In *RimWorld Wiki*. Retrieved October 3, 2024, from https://rimworldwiki.com/wiki/AI_Storytellers
28. Figure 11. Oterhals, J. (2022). Why does a 5,431-character story about Atari's 2 KB game Pong need 3,083? *Medium*. Retrieved from https://jcoterhals.medium.com/why-does-a-5431-character-story-about-ataris-2-kb-game-pong-need-3-08-30db27cd2648
29. Figure 12. Wikipedia contributors. (n.d.). M.U.L.E. Atari 8-bit PAL screenshot [Digital image]. *Wikipedia*. Retrieved October 3, 2024, from https://en.wikipedia.org/wiki/File:M.U.L.E._Atari_8-bit_PAL_screenshot.png

QUANTUM NARRATIVES

30. The Fibonacci sequence is a series of numbers where each number is the sum of the two preceding ones, starting with 0 and 1. The sequence begins: 0, 1, 1, 2, 3, 5, 8, 13, 21, 34...
31. Figure 14. Stockcake. (n.d.). Intricate geometric pattern [Digital image]. Retrieved from https://stockcake.com/i/intricate-geometric-pattern_1479789_1161529
32. Fronczak, A., Fronczak, P., Samsel, M. J., Makulski, K., Łepek, M., & Mrowiński, M. (2024). Scaling theory of fractal complex networks. *Scientific Reports*, 14, 9079. https://doi.org/10.1038/s41598-024-59765-2
33. Cohn-Sheehy, B. I., Delarazan, A. I., Reagh, Z. M., Crivelli-Decker, J. E., Kim, K., Barnett, A. J., Zacks, J. M., & Ranganath, C. (2021). The hippocampus constructs narrative memories across distant events. *Current Biology*, 31(22), 4935–4945.e7. https://doi.org/10.1016/j.cub.2021.09.013
34. Kurzweil, R. (2012). *How to Create a Mind: The Secret of Human Thought Revealed*. Penguin Books.
35. Herman, D. (2009). *Basic Elements of Narrative*. John WIley & Sons.

36. Oehrn, C. R., Hanslmayr, S., Fell, J., Deuker, L., Kremers, N. A., Do Lam, A. T., Elger, C. E., & Axmacher, N. (2014). Neural communication patterns underlying conflict detection, resolution, and adaptation. *Journal of Neuroscience, 34*(31), 10438–10452. https://doi.org/10.1523/JNEUROSCI.3099-13.2014
37. Oliver, C.A. (2023). The Social Brain and the Neuroscience of Storytelling. In: Rowland, S., Kuchel, L. (eds) *Teaching Science Students to Communicate: A Practical Guide*. Springer, Cham. https://doi.org/10.1007/978-3-030-91628-2_4
38. Zak P. J. (2015). Why inspiring stories make us react: the neuroscience of narrative. *Cerebrum : the Dana forum on brain science, 2015*, 2.
39. Rizzolatti, G., & Craighero, L. (2004). The mirror-neuron system. *Annual Review of Neuroscience, 27*(1), 169–192. https://doi.org/10.1146/annurev.neuro.27.070203.144230
40. Zak, P. J. (2013, December 17). How stories change the brain. *Greater Good Magazine*. Retrieved October 3, 2024, from https://greatergood.berkeley.edu/article/item/how_stories_change_brain
41. Fugate, J. M. B., Macrine, S. L., & Hernandez-Cuevas, E. M. (2024). Therapeutic potential of embodied cognition for clinical psychotherapies: From theory to practice. *Cognitive Therapy and Research, 48*, 574–598. https://doi.org/10.1007/s10608-024-10468-y
42. Potter, M. C., Wyble, B., Hagmann, C. E., & McCourt, E. S. (2014). Detecting meaning in RSVP at 13 ms per picture. *Attention, Perception, & Psychophysics, 76*(2), 270–279. https://doi.org/10.3758/s13414-013-0605-z
43. Raichle, M. E., MacLeod, A. M., Snyder, A. Z., Powers, W. J., Gusnard, D. A., & Shulman, G. L. (2001). A default mode of brain function. *Proceedings of the National Academy of Sciences, 98*(2), 676–682. https://doi.org/10.1073/pnas.98.2.676
44. Andrews-Hanna, J. R., Smallwood, J., & Spreng, R. N. (2014). The default network and self-generated thought: Component processes, dynamic control, and clinical relevance. *Annals of the New York Academy of Sciences, 1316*(1), 29–52. https://doi.org/10.1111/nyas.12360

45. Tornow, J. (2019, January 10). Cirque du Soleil and the neuroscience of awe: How the "Lab of Misfits" is using science to make us gasp. *Vox*. Retrieved from https://www.vox.com/culture/2019/1/10/18102701/cirque-du-soleil-lab-of-misfits-neuroscience-awe
46. Awh, E., Vogel, E. K., & Oh, S. H. (2006). Interactions between attention and working memory. *Cognitive Research: Principles and Implications*, 1(1), 3. https://doi.org/10.1186/s41235-016-0029-0
47. Shimamura, A. P. (2020, July). Psychocinematics and the language of film. *Psychology Today*. Retrieved from https://www.psychologytoday.com/us/blog/in-the-brain-the-beholder/202007/psychocinematics-and-the-language-film

INDEX

A

Aarseth, Espen 17. *See also* Interactive fiction
Accessibility 87. *See also* Inclusivity
Actors, 41–42, 51, 55, 73, 81
The Adaptive Prop 89. *See also* Thought experiments
Agency 17–21, 26, 29, 43, 45–46, 49, 80–81, 88–90, 94. *See also* Immersive theater; Live action role-playing (LARP); Murray, Dr. Janet; Narrative paradox; Participatory storytelling; Player agency Alternate Reality Games (ARG), 42. *See also* Games
Amygdala 64–65. *See also* Conflict
Anterior Cingulate Cortex 64
Apophenia 67
Archetypes 3, 38
Aristotle 2–4, 8, 10–13, 17–18, 23–26, 48–49, 55, 61–62, 67, 72, 74, 80. *See also* Catharsis; Classical storytelling; Four Causes; *Poetics*
Artificial Intelligence (AI) ii, 8, 19, 27–31, 37–39, 50, 59, 86, 88
AI-driven story structures 28. *See also* Dynamic content generation; Mateas, Michael; Young, Robert Michael; Riedl, Dr. Mark; Stern, Andrew
Artificial Intelligence ethicists 88
ChatGPT4 99
Augmented Reality (AR) 41, 44–46. *See also* Pokémon Go
Authorship 82
Avengers campus 42. *See also* Marvel Cinematic Universe
Avengers: Age of Ultron 42. *See also* Marvel Cinematic Universe
Aylett, Ruth 18. *See also* Narrative paradox; Louchart, Sandy

B

Babylonian creation story 22
Bates, Dr. Joseph 27–28, 31–32. *See also* Believable agents; Drama manager; Oz Project
Beat Saber 41. *See also* Virtual Reality (VR)
Beck, Julian 40. *See also* Living Theater
Behavioral economics 60. *See also* Patterns

Behavioral indicators 68. *See also* Engagement
Believable agents 27, 32, 88. *See also* Bates, Dr. Joseph; Drama manager; Oz Project
Bharata Muni 61. *See also* Indian classical drama; Rasa
Big Five personality traits 30. *See also* Crawford, Chris; Psychology
Black box theater 97–98
Black Mirror: Bandersnatch 19, 37
Blackbody radiation experiment 34
Boal, Augusto 73. *See also* Dialogue; Theatre of the Oppressed
Bogost, Ian 48. *See also* Complex systems; Procedural rhetoric
Bounded choices 42. *See also* Spatial narrative
Brain-computer interfaces 70
Branching narrative 16, 19, 36. *See also Black Mirror: Bandersnatch*; *Choose Your Own Adventure*; *Fahrenheit*
Breaking Bad 12
Brook, Peter 73. *See also The Empty Space*

C

Campbell, Joseph 11. *See also* Hero's Journey; Solitary heroism
Carnegie Mellon University 27. *See also* Bates, Dr. Joseph; Oz Project
Catharsis 2, 11–13, 24, 49, 62, 65, 77. *See also* Freytag's Pyramid; Tragedy; White, Walter

Chinese storytelling 66. *See also* Kishōtenketsu
Choose Your Own Adventure 15, 36
Cirque du Soleil 62, 69
Classical laws of physics 5, 24, 34, 58, 75. *See also* Newton, Isaac
Classical physics 33, 54–55
Classical storytelling 18, 36, 55–56, 86–87, 93
Climax 1, 11, 65, 77. *See also* Freytag's Pyramid
Cognitive load 43. *See also* Sweller, John
Colossal Cave Adventure 15. *See also* Text-based adventure games
Complex systems 25–26, 47–48, 86
Conflict 11, 13, 18, 54, 58, 64–66, 79, 94, 98
Copernicus, Nicolaus 4–5. *See also* Heliocentrism
Cortisol 65. *See also* Climax; Neurotransmitters
Crawford, Chris 30. *See also* Game design; Interactive storytelling
Cross-platform stories 42. *See also* Distributed narratives

D

Default Mode Network (DMN) 67. *See also* Immersion
Deterministic 54–56, 86, 108
Dialogue 12, 20, 43, 51–52, 73, 79, 88, 107–108
Diction 10, 12, 72. *See also* Aristotle; *Poetics*

Dinner theater 38
Distributed narratives 41. *See also* Cross-platform stories; Harry Potter
Divers 81. *See also* Skimmers; Swimmers
Dopamine 64–65. *See also* Neurotransmitters; Anterior Cingulate Cortex; Neurotransmitters
Double-slit experiment 34–35, 48
Drama manager 27. *See also* Mateas, Michael; Stern, Andrew; Tabletop role-playing game (TTRPG)
Dramatic arc 65, 71. *See also* Cortisol; Freytag's Pyramid; Narrative structure; Oxytocin
Dramatic flow 77, 79–80. *See also* Spacetime; Foundational components of quantum narratives
Dramatic narrative 25, 58, 72–73, 100–101
Dual-character mask 90. *See also* Thought experiments
Dungeons & Dragons 26. *See also* Tabletop role-playing game (TTRPG)
DuVernay, Ava 62
Dynamic content generation 36–37, 39, 44. *See also* Procedural generation
Dynamics 78–79. *See also* Fields; Foundational components of quantum narratives;

Foundational components of quantum narratives

E

Einstein, Albert 7, 25, 53–54, 57–58, 74, 103. *See also* Quantum physics; Subatomic; Time dilation
Electroencephalography (EEG) 69
Electromagnetic fields 58
Electric fields 77
Emergent gameplay 18, 20
Emotion-focused therapy 65. *See also* Catharsis
Emotional resonance 47, 60
The Empty Space 73. *See also* Brook, Peter
En Garde Arts 40. *See also* Hamburger, Anne
Energy 25, 34, 58, 74–75, 77–79, 101. *See also* Foundational components of physics
Engagement 11, 16, 18–19, 22, 27, 29, 43, 51, 56, 61–63, 66, 68–73, 77–79, 81–83, 86, 88, 104. *See also* Neural engagement
Environment 12, 20, 38, 40–41, 43, 46, 51, 56, 65–66, 69, 77–79, 81–82, 92–93, 97–98, 108. *See also* Setting
Environmental storytelling 87
Environmental theater 40. *See also* Schechner, Richard
Ethologists 88
Euclid 24
Event schemas, 64. *See also*

INDEX

Psychology
Everything, Everywhere, All At Once 12
Evolution over time 105. *See also* Quantum postulates
Experiential marketing 45. *See also* Interactivity
Exposition 10. *See also* Freytag's Pyramid

F

Façade 19, 28, 31–32. *See also* Interactive drama; Stern, Andrew
Fahrenheit 36. *See also* Branching narrative; Games
Falling action 11. *See also* Freytag's Pyramid
Fernández-Vara, Clara 18. *See also* Game studies; Narrative paradox; Storytelling
Fibonacci sequence 59
Fields iii, 17, 27, 30, 58, 74–75, 77–79, 97. *See also* Dynamics; Foundational components of physics
Film (Filmmaking) i, 1, 19, 41–42, 47, 56, 63, 73, 83
Forces 22, 25, 30, 49, 56, 58, 73–75, 77–78, 95, 98. *See also* Theme
Foundational components of physics 74. *See also* Energy; Forces; Matter; Spacetime
Foundational components of quantum narratives 75. *See also* Dramatic flow; Dynamics; Guest (definitions of); Sensory forms; Theme
Four Causes 3–4, 8, 17, 23. *See also* Aristotle; Laurel, Dr. Brenda
Fractal-based algorithms 38. *See also No Man's Sky*
Fractals 59
Freytag's Pyramid 10, 64
Freytag, Gustav 10–11, 64
Functional magnetic resonance imaging (fMRI) 69

G

Gaiman, Neil 92
Galilei, Galileo 5, 23
Game design 17–18, 30
Game engines ii, 86
Game mechanics 46
Gameplay mechanics 16, 98
Game studies 20. *See also* Ludology
Game theorists 88
Games ii, 13, 15–22, 26, 30–31, 36–39, 41–42, 44–46, 48–49, 60, 86, 88, 98, 103. *See also Beat Saber*; *Colossal Cave Adventure*; *Half-Life*; *Halo*; *Hamlet*; *M.U.L.E.*; *Marvel Future Revolution*; *Mass Effect*; *No Man's Sky*; *Rogue*; *Zork*
General Relativity 7, 53. *See also* Einstein, Albert
Georgia Institute of Technology 29
Goethe, Johann Wolfgang 11. *See also* Werther fever
Gravity 53–54, 58, 75, 87
Guest (definitions of) 79–83.

See also Agency; Energy;
Participants; Players; Users;
Foundational components of
quantum narratives

H
Half-Life 16
Halo 41
Hamlet 17, 62, 67, 92. See
  also Classical storytelling;
  Shakespeare, William
Haptics, 46
Harry Potter 41. See also
  Distributed narratives
Hawking, Stephen 8
Heliocentrism 4. See also
  Copernicus, Nicolaus
Hero's Journey 11
Hippocampus 63
Humanity ii–iii, 26, 85, 93, 95–96

I
Immersion 27, 70, 87
Immersive theater 20, 40, 43. See
  also Live theater
Improv theater 38. See also Dinner
  theater
Inclusivity 87. See also Accessibility
Index of Forbidden Books 5. See
  also Copernicus, Nicolaus
Indian classical drama 61, 66. See
  also Natyashastra
Intensified continuity 83. See also
  Film (Filmmaking)
Interactive acting 81. See also
  Wirth, Jeff

Interactive drama 18–19, 21–22,
  28, 31. See also Façade
Interactive fiction 17–18, 44. See
  also Aarseth, Espen
Interactive media 17, 19, 21,
  28, 32, 48. See also Juul, Jesper;
  Millard, David
Interactive narrative 17–18, 22,
  27, 30, 32, 48–49, 56, 74–75,
  86–88. See also Complex systems;
  Murray, Dr. Janet
Interactive storytelling 17, 29–31,
  45, 47, 88. See also Laurel, Dr.
  Brenda
Interactivity i, 15, 18, 20, 36,
  44–45, 47–49, 60, 73, 80, 86, 108
Internet of Things (IoT) 46–47
Intertwined plotlines 12
Isbister, Katherine 20
Italian Futurists 40

J
Japanese storytelling 61, 66. See
  also Kishōtenketsu; Motokiyo,
  Zeami
Jenkins, Henry 92
John, Elton 63
Juul, Jesper 21. See also Interactive
  media

K
Kael, Pauline 73. See also
  Emotional resonance
Kepler, Johannes 23. See also Laws
  of planetary motion
Kishōtenketsu 66. See also Chinese

storytelling; Japanese storytelling
Kuhn, Thomas, 57

L

Laurel, Dr. Brenda 17–18. *See also* Interactive storytelling; Four Causes
Laws of planetary motion 23. *See also* Kepler, Johannes
Leading question bias 68. *See also* Self-reported surveys
Linear i, 12, 18, 20, 22–23, 40, 48–51, 56, 60, 66, 72–73, 93, 108
The Lion King 63. *See also* Taymor, Julie
Live action role-playing (LARP) 26–27, 38–39
Live theater 20. *See also* Black box theater; Immersive theater
Live theatrical virtual reality 19. *See also* Tender Claws; *The Under Presents*; *The Under Presents*
Living Theater 40. *See also* Beck, Julian
Louchart, Sandy 18. *See also* Aylett, Ruth
Ludology 20–21. *See also* Game studies
Ludonarrative 19. *See also* Roth, Christian

M

M.U.L.E. 44–45
Macbeth 20. *See also* Classical storytelling; Shakespeare, William
Marvel Cinematic Universe 42, 95. *See also* Avengers campus; *Avengers: Age of Ultron*; *Marvel Future Revolution*; *WandaVision*
Marvel Future Revolution 42. *See also* Marvel Cinematic Universe
Mass Effect 16
Mateas, Michael 18–19, 27–29. *See also Façade*; Interactive drama; Stern, Andrew
Mathematics ii, 7–9, 24–25, 57–58
The Matrix 63
Matter 23, 33, 36, 51, 58, 68, 72, 74–75, 77–79, 82, 93, 95. *See also* Sensory forms
McKee, Robert 73. *See also* Screenwriting
Measurement values 100. *See also* Quantum postulates
Media studies 21, 88
Melody 10, 12–13, 72. *See also* Aristotle; *Poetics*
Millard, David 19. *See also* Interactive media
Miller, Robyn 16. *See also Myst*
Mimesis 55. *See also* Aristotle
Minecraft 37. *See also* Procedural generation; Open-world games
Mirror neurons 65. *See also* Narrative transportation
Mixed Reality (MR) 41
Moral molecule 65. *See also* Zak, Dr. Paul J.
Motokiyo, Zeami 61. *See also* Japanese storytelling
Multi-platform experiences 44

Murray, Dr. Janet 17–18, 21. *See also* Interactive narrative
Museums 45
Mutual evolution 83
Myst 16, 45. *See also* Games; Miller, Robyn
Myth 21–22, 94

N

Narrative engine 98–100, 102–104, 108
Narrative paradigm ii, 93, 95
Narrative paradox 15, 17–21, 23, 25, 27, 29, 31–32, 46–48, 56. *See also* Aylett, Ruth; Louchart, Sandy; Murray, Dr. Janet
Narrative planning model 29
Narrative structure 16–19, 25, 66
 Narrative arc 18, 20, 28, 43, 46, 64
Narrative systems 26–27, 30, 88
Narrative transportation 65. *See also* Mirror neurons
Natyashastra 61-62
Neocortex 63
Nerve signal transmission 58
Network theory 59
Neural engagement 68
Neuroscience 63–66, 68, 70, 72–73, 81, 86, 90, 98, 102, 105
 Neurochemical 65, 70
 Neuroscientists 69, 88
 Neurotransmitters 64. *See also* Dopamine
Newton, Isaac 5-6. *See also Principia Mathematica*

Newtonian physics 53–54. *See also* Classical physics
The Night of January 16th 15. *See also* Live theater; Rand, Ayn
No Man's Sky 37–38. *See also* Fractal-based algorithms; Procedural generation
Not Tonight 49. *See also* Games
Nuclear fusion 58. *See also* Electron

O

Observables 101. *See also* Quantum postulates
Open-world games 16–17. *See also Red Dead Redemption 2*; *The Sims*; *Minecraft*
Orwell, George 49, 61
Overlapping narratives 42. *See also* Spatial narrative
Oxytocin 65, 70. *See also* Moral molecule; Neurotransmitters
Oz Project 27–28, 31–32

P

Papers, Please 49. *See also* Pope, Lucas; Procedural rhetoric
Participants 20, 22, 43, 46, 69–70, 81–82
Participatory storytelling 18, 41. *See also* Interactive storytelling; Participants
Patterns 35–37, 48, 51, 55, 58–60, 63–64, 67–69, 86. *See also* Event schemas; Social networks
Physics ii, 2, 8, 23–24, 33–34, 36,

44, 53–55, 57, 73–79, 86, 98
Player agency 17–19, 26, 29, 45, 49
Player choice 17, 26, 49, 108
Player freedom 19–20
Players 16, 18, 20, 26–27, 31, 38–39, 41, 44, 48, 60, 81
Pleasure-Arousal-Dominance (PAD) 31
Plot 1–2, 10–13, 15, 17, 21, 23–24, 26, 28–29, 32, 36, 39, 43, 48–51, 56–58, 63–64, 66–68, 72–73, 75, 77–79, 83, 88, 90–91, 96, 104, 107
Mythos 10. *See also* Aristotle
Poetics 2, 13, 17, 23, 25, 32, 47, 62, 74, 80. *See also* Aristotle; Catharsis; Classical storytelling; Traditional plot structure; Western drama; Western dramatic theory
Pokémon Go 41. *See also* Augmented Reality (AR)
Pong 44
Pope, Lucas 49. *See also Papers, Please*
Popper, Karl 57
Posterior Cingulate Cortex 69
Prefrontal Cortex 64, 69
Principia Mathematica 5–6. *See also* Newton, Isaac
Probabilistic 54–56, 86, 98, 102
Probability 54, 74, 78, 91, 108
Probability and measurement 102. *See also* Quantum postulates
Procedural generation 37–38.
See also Dynamic content generation; *Minecraft*; *No Man's Sky*; *Rogue*
Procedural rhetoric 48–49. *See also Papers, Please*
Protagonist 1, 11, 49, 93–95
Psychology 17, 30, 57–58
Behavioral psychology 88. *See also* Event schemas
Punchdrunk 20. *See also Sleep No More*
Pythagorean theorem 24, 47. *See also* Euclid

Q

Quantum narrative ii, 2, 8, 10, 12, 16, 18, 20, 22, 24, 26, 28, 30, 32, 34, 36, 38, 40, 42, 44, 46, 48, 50, 52–83, 86, 88–90, 92–93, 96, 98, 100, 102, 104, 106, 108
Quantum physics 36. *See also* Einstein, Albert
Quantum mechanics ii, 54–56, 95, 99–102, 104, 108
Quantum postulates 98
Quantum spin 106–107
Quantum sand timer 91. *See also* Thought experiments

R

Rand, Ayn 15
Rapid processing power 66
Rasa 61–62
Rasa theory 66
Reactive characters 42. *See also* Spatial narrative

Recall bias 68. *See also* Self-reported surveys
Red Dead Redemption 2 17. *See also* Open-world games
Relative linearity 71
Resolution 1, 11–12, 32, 36, 64–66, 71, 77. *See also* Freytag's Pyramid
Response bias 68. *See also* Self-reported surveys
Rhythm 13, 41, 63. *See also* Aristotle; *Poetics*
Riedl, Dr. Mark 29. *See also* Georgia Institute of Technology; Narrative planning model; *Scheherazade*
RimWorld 38. *See also* Games; Procedural generation
The Rippled Mirror 91. *See also* Thought experiments
Rising action 10–11. *See also* Freytag's Pyramid
Rogue 37. *See also* Procedural generation
Roth, Christian 19. *See also* Ludonarrative; Ludonarrative
Ryan, Maria-Laure 21. *See also* Games

S

Schechner, Richard 40. *See also* Environmental theater; Environmental theater
Scheherazade 32. *See also* AI-driven story structures; Riedl, Dr. Mark
Screenwriting 11, 73. *See also* McKee, Robert
Self-reported surveys 68. *See also* Leading question bias; Recall bias; Response bias
Sensory forms 79. *See also* Matter; Foundational components of quantum narratives
Setting 1–3, 7, 9, 11, 13, 26, 51–52, 65, 69, 74–75, 77, 100–102
Shakespeare, William 13, 20, 67. *See also* Classical storytelling; *Hamlet*; *Macbeth*
The Sims 16, 31. *See also* Open-world games
Site-specific theater 40. *See also J.P. Morgan Saves the Nation*; En Garde Arts
Hamburger, Anne 40. *See also J.P. Morgan Saves the Nation*; En Garde Arts
J.P. Morgan Saves the Nation 40. *See also* Hamburger, Anne
Skimmers 81–82. *See also* Divers; Swimmers
Skin conductivity 69. *See also* Cirque du Soleil; Electroencephalography (EEG)
Sleep No More 20, 43. *See also* Punchdrunk
Social cuing 65. *See also* Moral molecule; Zak, Dr. Paul J.
Social networks 59. *See also* Patterns
Social neural networks 67
Social VR 41. *See also* Virtual

Reality (VR); VRChat
Society 21–22, 25, 56, 82, 94
Sociolinguists 88
Solitary heroism 95
The Sorrows of Young Werther 11. See also Goethe, Johann Wolfgang; Werther fever
Spacetime 8, 54, 74–79. See also Dramatic flow
Spatial narrative 20, 36, 39, 43, 81
Spatial theater 40, 42–43
Spectacle 10, 12–13, 20–21, 62, 72. See also Aristotle; *Poetics*
Spielberg, Steven 92
State of the System 99–100. See also Quantum postulates
Stern, Andrew 19, 28. See also *Façade*; Mateas, Michael
Storytelling i–ii, 2, 4, 11–12, 15, 17–22, 24, 27, 29–31, 36, 38–39, 41–42, 44–50, 56, 66, 71, 81–82, 86–89, 93–94, 98, 108
Story structures 10, 28, 64, 66
Storytelling process 12
Subatomic 33–34, 54, 56, 75–76, 85. See also Einstein, Albert
Electron 33, 36, 58, 99. See also Nuclear fusion
Experiments 34, 89
Neuron 65
Proton 33
Sweller, John 43. See also Cognitive load
Swimmers 81. See also Divers; Skimmers

T

Tabletop role-playing game (TTRPG) 26. See also *Dungeons & Dragons*
Taymor, Julie 63. See also *The Lion King*
Technology i–iii, 2, 15, 17, 26, 33, 36–37, 40, 46–47, 57, 70, 93
Teleological 3
Telescope 5, 85
Tender Claws 19. See also *The Under Presents*
Text-based adventure games 15, 44. See also Games; *Colossal Cave Adventure*; *Zork*
Theatre of the Oppressed 73. See also Boal, Augusto
Theme 12, 41, 49, 75–78, 90. See also Forces; Foundational components of quantum narratives
Then She Fell 20, 43. See also Immersive theater; Third Rail Projects
Third Rail Projects 20
Thought 10, 12, 72. See also Aristotle; *Poetics*; Theme
Thought experiments, 89–91. See also Dual-character mask; Quantum sand timer; The Adaptive Prop; The Rippled Mirror
Three Unities 2, 10. See also Aristotle
Three-act structure 11. See also Story structures; Traditional

plot structure; Western story structure
Time dilation 53. *See also* Einstein, Albert
Traditional narrative 17, 20, 49, 64
　Traditional plot structure 17, 83
　Traditional storytelling 42, 44
Tragedy 2–3, 10, 61, 67. *See also* Catharsis
Twitch 83

U

Ultraviolet catastrophe 34. *See also* Blackbody radiation experiment
The Under Presents 19. *See also* Live theatrical virtual reality; Tender Claws
Users 29, 41, 46, 81

V

Virtual Reality (VR) i–ii, 19, 40–41, 46, 80, 86. *See also* Beat Saber; Live theatrical virtual reality; Social VR
VRChat 41. *See also* Social VR

W

WandaVision 42. *See also* Marvel Cinematic Universe
Wave function collapse 103–104. *See also* Quantum mechanics; Quantum postulates
Werther fever 11. *See also* Goethe, Johann Wolfgang
West African storytelling 66

Western drama 1–2, 4–5, 11, 13, 23, 25, 63, 65–66, 94
Western dramatic theory 2
Western story structure 11, 65
Westworld 12
White, Walter 12. *See also Breaking Bad*; Catharsis
Wirth, Jeff 81. *See also* Interactive acting

Y

Yoruba oral storytelling 66. *See also* West African storytelling
Young, Robert Michael 29. *See also* Narrative planning model; Riedl, Dr. Mark
YouTube 83

Z

Zak, Dr. Paul J. 64–65, 70. *See also* Moral molecule; Neuroscientists; Oxytocin
Zimmer, Hans 63. *See also* Melody
Zombies, Run! 46
Zork 15. *See also* Text-based adventure games

www.ingramcontent.com/pod-product-compliance
Lightning Source LLC
Chambersburg PA
CBHW072151200426
43209CB00052B/1114